T0222176

PV Technical Sales

NABCEP sets the standard for solar certifications in the United States and Canada. The NABCEP PV Technical Sales Certification shows customers, friends and employers that you are knowledgeable and qualified to sell solar systems.

If someone is selling solar, they need to know what they are selling and how it is configured. Where will they connect the circuit breaker? Will the house need expensive modifications in order for a PV system to be installed? These are the questions that you as an NABCEP Technical Sales Certified solar salesperson will confidently answer.

This book is full of practical information that anyone selling solar should know in order to properly serve their customers and to ethically represent the industry that is solving the world's problems on the ground and rooftop level.

This book will be of use to those taking the NABCEP PV Technical Sales Exam, as well as anyone selling or planning to sell solar.

Sean White is an experienced teacher, instructor and solar PV professional. He received the 2014 Interstate Renewable Energy Council Clean Energy Trainer of the Year Award at Solar Power International. Sean teaches classes for many organizations around the world.

PV Technical Sales

PREPARATION FOR THE NABCEP
TECHNICAL SALES CERTIFICATION

Sean White

Routledge
Taylor & Francis Group

LONDON AND NEW YORK

from Routledge

First published 2016 by Routledge

Published 2022 by Routledge
4 Park Square, Milton Park, Abingdon, Oxon OX14 4RN
605 Third Avenue, New York, NY 10017

Routledge is an imprint of the Taylor & Francis Group, an informa business

British Library Cataloguing-in-Publication Data
A catalogue record for this book is available from the British Library

Library of Congress Cataloging-in-Publication Data
White, Sean (Electrical engineer), author.
    PV technical sales : preparation for the NABCEP technical sales
certification / Sean White.
        pages cm
    Includes bibliographical references and index.
    1. Photovoltaic power systems—United States—Costs—
Examinations—Study guides. 2. Buildings—Electric equipment—
United States—Costs—Examinations—Study guides. 3. Buildings—
Energy consumption—United States—Examinations—Study guides.
I. Title.
    TK1087.W4485 2016
    621.31'24076—dc23   2015031610

ISBN: 978-0-415-71334-4 (pbk)
ISBN: 978-1-315-77008-6 (ebk)

Typeset in Rotis Sans Serif
by Keystroke, Station Road, Codsall, Wolverhampton

Contents

Preface

Solar systems are often sold by great salespeople that have little to no knowledge of what they are selling. This is a problem that has a solution – studying this book. Another benefit of reading this book is getting the ability to pass the NABCEP Technical Sales Certification Exam, which is by far the easiest way to become NABCEP Certified.

A car salesperson can sell a car without knowing the first thing about how to build or even drive a car, but to sell a solar system, a salesperson should know how to configure a PV system. They need to know how a PV system that will work great in Hawaii could be a fire hazard in Alaska due to increased voltage at cold temperatures.

When a solar salesperson is at the customer's home, they should know how the PV system is going to be connected to the utility. A breaker size that is different could mean a $1000 difference in the cost of the system.

If PV modules are placed on a roof differently with regards to shading, it could be a big problem that would have repercussions greater than $1000 over the life of the system. Oftentimes, wiring the system in a different manner will make a big difference when it comes to system production.

Since the solar PV market is so new and the growth rate is beyond exponential, many exceptional salespeople have transitioned from another career into the fast-growing and lucrative solar market. Understanding how solar systems work is paramount in getting a happy customer or a happy employer. Many times solar systems have been sold that would not operate correctly or were extra expensive to install, because the salesperson did not properly understand the technical aspects of photovoltaics.

A PV system can be installed to be the same visually as another system across the street, but can perform drastically differently because of how the system is wired underneath the PV modules, intermittent shading, whether the inverter fits the application or if the system fits the utility rate schedule. Inexperienced solar salespeople have often tried to fit as many PV modules on the roof as they can or tried to offset 100% of every bill. Strategically, there are many reasons that a professional solar PV salesperson must have technical knowledge in order to best serve his or her customer and generate repeat business and a good reputation.

Employers will value most a salesperson who on their own can generate sound proposals that will make the most money for the company and the customer. If a salesperson can understand how the system will connect to the existing electrical service, they will make better-informed choices regarding the cost of the system. If a salesperson understands how the system is installed and engineered and they can sell, then they will be the most valuable person in the company.

This book will give the reader the knowledge required for solar PV salespeople to be successful and of the greatest value to their customers, their companies and themselves. An excellent way of demonstrating your knowledge is to become NABCEP Certified, and the most productive way to become NABCEP Certified is by reading this book, studying for and passing the NABCEP Solar PV Technical Sales Exam.

If you would like to get every NABCEP PV exam under your belt in order of difficulty, first get warmed up by taking the NABCEP PV Entry Level Exam, then take the NABCEP PV Technical Sales Exam, and after that take the famously difficult NABCEP PV Installation Professional Exam. Collecting all three in order is a good progression and will set you apart from others.

Since much of the material covered on the NABCEP Entry Level Exam is also covered on the NABCEP PV Technical Sales Exam, it is recommended to take and pass the Entry Level Exam and/or to read the book *Solar Photovoltaic Basics*, which is the first book in this series written to cover the basics that everyone taking any NABCEP exam should know. To pass the Technical Sales Exam with flying colors, also study the book *Solar PV Engineering and Installation*, by the same author.

This book is written not only to best prepare you to understand the technical aspects required to be a great solar PV salesperson, but to also assist you in studying the right material in order to most efficiently pass the NABCEP Technical Sales Certification Exam.

Best of luck!

Sean White

PV technical sales: facts and figures

To make the most of your time, rather than digging deep into big paragraphs with a highlighter for important points, here is some material that is ripe for memorization or flash cards, which includes many acronyms.

NREL (National Renewable Energy Lab) Formerly the Solar Energy Research Institute, NREL is based in Golden, Colorado. Much of the research in the world on solar PV has been done at NREL. NREL also has many websites, including PVWATTS (pvwatts.nrel.gov) and System Advisor Model (sam.nrel.gov).

PVWATTS An NREL website that is used to predict output of PV systems. PVWATTS is very simple to use and is the most commonly used website for sizing PV systems. Other software that is used to predict outputs and do financial analysis of PV systems often uses information from PVWATTS.

SAM System Advisor Model is an NREL-designed software that can predict system outputs and will do financial analysis for PV and other renewable energy systems. SAM is more detailed and complicated to use than PVWATTS and can get information from different utility databases.

NREL Redbook Thirty years' worth of solar radiation data for locations in the United States. This is a good place to compare the solar radiation as expressed in kWh/m²/day at flat tilt, latitude tilt, latitude +15 tilt and latitude −15 tilt for many locations.

FIT (feed-in tariff) A type of PBI. Typically a FIT is a 20-year fixed price for electricity produced by solar. The government requires the utility to pay the price. Introduced in Germany.

TOU (time of use) A utility rate schedule where the price of energy changes based on time. Typically weekday afternoons are most expensive due to an increased demand for electricity. This is why people that have a TOU schedule or expect to have a TOU schedule in the future would be better off facing their PV modules west than east. Some people expect that in the future, with electric cars, everyone will have a TOU rate schedule.

Tiered rate schedule A rate schedule where rates go up as more energy is used. Tiered rates are typically residential and based on monthly energy usage. Tiered rates penalize people for using a lot of energy; however, they benefit high energy users disproportionally for getting solar.

Derating factor A number that you multiply by to compute for losses. Often common "dc to ac" derating factors are in the range of 0.77 to 0.85.

In many places, including the NREL website, the derating factor is termed "dc to ac," but it is actually a factor that will convert kWh per square meter per day to ac kWh per kW of PV per day. Sometimes we are given a loss factor and we would subtract that from 1 to get a derating factor. For instance, if we lose 20%, our loss factor would be 0.2 and our derating factor is going to be 0.8. The NREL PVWATTS calculator has transitioned from a derating factor to a loss factor.

PPA (power purchase agreement) An agreement to purchase electricity for a specified time, often about 20 years. As opposed to a lease, with a PPA you are paying for the amount of energy the system produces. A PPA typically has a buyout at the end and may have early buyout provisions. The owner of the system who sells the energy will take advantage of the tax credits, depreciation and other incentives.

10W/square foot An estimate that some people use to quickly determine how much PV will fit on a roof. This estimate is very conservative and in some cases it can be possible to get 20W/square foot with efficient PV. 10W/square foot can account for unused roof space, such as fire setbacks, other equipment on roofs and pathways.

SOH CAH TOA Used to memorize trigonometry, which can be used to calculate roof slopes and shadow lengths. Sine = opposite/hypotenuse; cosine = adjacent/hypotenuse; tangent = opposite/adjacent. Trigonometry uses relationships of right triangles to determine angles and side lengths of the triangle. (There is an entire chapter on solar trigonometry in the second book in this series, *Solar PV Engineering and Installation* and part of a chapter on trigonometry in this book.)

Voltage Electrical pressure. The dangers with higher voltage is arcing and sparking. Lightning is very high voltage. The hydraulic analogy for voltage is pressure.

Current Electrical flow. The dangers with high current are heat and stoppage of the heart. Current is what kills.

Power Voltage multiplied by current is power. Power is the instantaneous rate of electricity usage. If a utility charges for power, it is called a demand charge. The basic fundamental unit of power in the solar industry is the watt (W) or the kilowatt (kW).

Demand charge Charges for peak power usage. Often at 15-minute intervals or instantaneously, power is measured and the highest power measurement of the month determines the demand charge. When everything is on at once, that will cause a higher demand charge. Solar without energy storage typically does *not* lower demand charges, since it is not always sunny. Incorporating energy storage into a PV system can help offset demand charges by storing energy when demand is low and using the stored energy when the demand is high.

Energy Energy is power multiplied by time. Energy is the quantity of electricity used and is the main thing that utilities charge for, in kWh. When we design our PV systems, we are usually most interested in the energy they will make over time.

Irradiance Solar power, which can be measured in watts per square meter. This is a unit of power per unit area.

Peak sun Irradiance of 1000 watts per square meter.

Irradiation Solar energy, which is often measured in kWh per square meter.

PSH (peak sun hours) Solar energy measured in kWh/square meter. Typically PSH are measured for a day for a particular tilt angle and are a measurement of insolation.

Insolation (incident solar radiation) Solar radiation measured at a particular tilt angle, such as latitude tilt, and measured in kWh/m²/day. Insolation, PSH and irradiation can all be the same thing. Insolation is not a measurement of ac kWh (electricity), it is sunlight hitting a square meter.

Solar module A single unit of PV that people will purchase. An average solar module is 250W, has 60 cells and is about 39 inches (1m) × 66 inches (1.67m). Solar modules are often incorrectly called solar panels. Almost all solar modules have all of the solar cells arranged in series.

60-cell module Most 60-cell solar modules are arranged in a 6 × 10 cell format and have bypass diodes that divide the cells electronically into three groups of 20 cells.

72-cell module Most 72-cell modules are arranged in groups of 6 × 12 cells and have bypass diodes that divide the cells electronically into three groups of 24 cells each.

Bypass diodes Bypass diodes can help current bypass shaded solar cells. When any cell in a group is shaded, the current will bypass the group of shaded cells. Bypassing the shaded cells will cause a decrease in voltage. If there were no bypass diodes, the current would have trouble bypassing the shaded cells and would create heat and resistance, which could be a fire hazard. When a bypass diode is broken, it will bypass cells, even without shade, and the module will typically see a one-third of a module decrease in voltage.

12-volt module In the early days of PV, modules had 36 cells and worked well for charging 12V batteries. We still see 12V modules for small battery charging systems and other small solar modules often have 36 cells.

Solar panel A group of solar modules connected together before installation. A solar panel technically is not a single solar module according to the National Electric Code, but most solar salespeople call a module a panel when speaking to their customers, since that is the common term that non-technical people use. For any technical exam, consider that a solar panel is a group of modules.

STC (standard test conditions) The way PV modules are tested, so that we will know their performance and can compare different modules to each other. The conditions are 1000W/m^2 irradiance, 25C (77F) cell temperature (not ambient) and 1.5 air mass. Modules in reality perform at conditions that are considerably below STC most of the time. For example, 5kW of PV will never export 5kW of electricity.

AM (air mass) The thickness of the atmosphere, which filters sunlight and determines the spectrum of sunlight. PV is tested at STC with an air mass of 1.5. In theory, if someone is at sea level and the sun is straight overhead, the air mass will be 1.5. Every day, due to humidity, pollution and other atmospheric conditions, the air mass will be different. Air mass was standardized in Cape Canaveral, Florida, on an equinox day at solar noon.

Wp (watts peak) When people are referring to an amount of PV, they often use the term watts peak, which means the number of watts at STC. Also, we can refer to kWp for kW at STC.

Series Connected together electrically positive to negative, which will increase voltage, but not current. A thousand solar cells connected together in series will have 1000 times the voltage of one solar cell and will have the same current as a single solar cell. With series connections, when the current goes through one device, it will go through all of the other devices that it is in series with.

Parallel Connected together so that the positive terminals are all connected and the negative terminals are also connected together. Parallel connections do not increase voltage and only increase current. Parallel connections are done at a combiner box. Also, parallel connections can be done with alternating current at a service panel or a sub panel.

PV source circuit PV modules connected in series. Often a PV source circuit is called a "string." PV modules or cells that are connected together in series will increase voltage.

PV output circuit Circuit at the output of a combiner box, often going to an inverter or charge controller. When PV source circuits are connected together in parallel at a combiner box, they will become a PV output circuit at the output of the combiner box.

Combiner box Where source circuits are combined to form a PV output circuit. Parallel connections are made at combiner boxes. Usually there will be fuses at a combiner box, unless only two strings are combined. (If strings are not combined, it is called a junction box or a transition box.)

IV curve A curve in which current (I) is plotted on the y-axis (vertical) and voltage (V) is plotted on the x-axis (horizontal). Typical points plotted on a PV IV curve are Isc (upper left), Voc (lower right), and Vmp and Imp, which are both at the middle "knee" of the curve. MPPT takes place at Imp and Vmp.

MPPT (maximum power point tracking) Optimizing power from a PV array, a PV source circuit or a PV module by operating at the best combination of current and voltage for maximum power. MPPT will operate at Vmp and Imp on an IV

Figure 1.1 IV curve, courtesy of Solmetric.

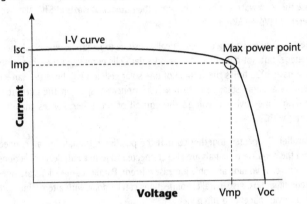

curve and will adjust to different environmental conditions, such as irradiance and temperature, for maximum power in the given situation.

Maximum power point tracker A part, often in an inverter, where MPPT takes place. Some inverters have two or more MPPT inputs and there should never be different PV source circuit lengths on a single maximum power point tracker.

Microinverter An inverter that operates underneath a single PV module. Microinverters are convenient, since they perform MPPT on each module individually. Microinverters can work with different module types, with different orientations; more modules can be added later and shading is less of an issue than with traditional string inverters. Microinverters are more expensive than string inverters.

Power optimizer Electronics underneath a module where dc to ac conversion and MPPT is done. Power optimizers are typically less expensive than microinverters, but they still need inverters to work with in order to produce ac power. Power optimizers can work with different module types, with different orientations; more modules can be added later and shading is less of an issue than with traditional string inverters.

Rafters The strong supporting parts of the roof, which the PV system is usually attached to. Rafters are underneath the roof deck and are oriented from the ridge top of the roof down to the lower eave of the roof. Wooden roofs often have rafters, which solar structures are attached to.

Purlins The strong supporting parts of the roof that go horizontally across the roof, from side to side. Purlins are usually what solar structures are structurally attached to on metal roofing systems.

Portrait Mounting modules with the long edges to the sides. Just like when you print portrait or a portrait painting in a museum. When the roof has rafters that are going from the ridge of the roof down, the rails are often mounted perpendicular to the rafters and the modules are usually mounted portrait with common rail systems.

Landscape Mounting modules with the short edges to the sides, like a land-scape painting in an art gallery. With roofs that have purlins that are parallel to the ridge of a roof, which is common with metal roofing systems, then the rails are often mounted perpendicular to the purlins and will come from the ridge of the roof down. These systems usually have modules mounted landscape. There are also railless systems that require modules to be mounted landscape on rafter roofs.

Building department The building department is usually run by a city or a county and typically must approve everything. You will go to the building department to apply for a permit, pay fees and ask for permission to make changes. The build-ing department will review your plans and permit package in order to determine whether you may install a PV system. Building departments will interpret the Code in their own manner and may have requirements that are not in the Code.

AHJ (authority having jurisdiction) The AHJ can be the building department, the inspector, the utility or any other entity that has authority over a project. It is the AHJs interpretation of the Code that is enforced and the AHJ can also enforce other things that are not in the Code.

NEC (National Electric Code) The rules for installing electrical systems for the United States and many other places. The NEC comes out with a new version every three years. The NEC is adopted by different states and jurisdictions at different times. The NEC is published by the National Fire Protection Association (NFPA) and the main goal of the NEC is to prevent fires. The NEC has the most comprehensive and detailed rules regarding solar PV installations of any official document in the world.

PE (professional engineer) A licensed engineer who can officially stamp drawings and approve a design. Different professional engineers are licensed in different states. A PE in one state cannot usually approve a project with a stamp in another state. Every state has different rules. Various types of engineers will need a PE stamp to officially approve plans.

Structural engineer An engineer who specializes in the viability and strength of a structure. A structural engineer is often consulted in areas of high winds, deep snow and for large buildings. Some building departments require a structural engineer for every PV installation on a building.

Electrical engineer An engineer who specializes in electrical systems and has gone to school to study electrical engineering. Electrical engineers are best suited to designing solar PV systems and are often on the design team for larger solar PV projects. They are especially useful when selecting transformers and determining interconnection points. Electrical engineers are well suited to interpreting the NEC. In some places in the NEC, design can be done under engineering supervision, such as interconnections done to center-fed panelboards (service panels).

Tilt angle The angle from horizontal that PV is tilted. Flat is a zero-degree tilt and vertical is a 90-degree tilt. Occasionally you will see a tilt angle that is from vertical; however, it is customary in the solar industry to cite tilt angles from the horizontal. A solar module with a 10-degree tilt angle would directly face the sun if the sun was at an 80-degree elevation angle.

Optimal tilt angle For winter latitude +15 degrees, for summer latitude −15 degrees; for year-round latitude 30 degrees or a tilt angle calculated with software.

Latitude +15 tilt Generally accepted as a good tilt angle to optimize winter performance of PV systems.

Latitude −15 tilt Generally accepted as a good tilt angle to optimize summer performance of PV systems.

30-degree tilt Generally accepted as a good tilt angle to optimize annual performance of PV systems in the continental United States.

Latitude tilt A good tilt angle for annual performance of PV systems for most of the world.

Low-tilt problem For framed modules at low tilt, there will be extra soiling problems. Less than 5 or 10 degrees is often considered a problem. The soiling will build up at the lower edge of the module, on the glass above the frame. Rain will take care of this problem by rinsing the dirt off of the module. In places with regular rain, this is not a problem. For modules without frames, there is not the same tendency for soiling build-up at the lower edge.

Power factor When voltage and current get out of phase, the power factor decreases. We often think of current and voltage being perfectly in phase; however, depending on the loads on the grid, power factor is often less than perfect. Typical utility interactive inverters have perfect power factor, but in some cases it is desirable for the utility to have inverters that can produce power that is less than perfect when there is a lot of solar on the grid. When current and voltage are perfectly in phase, then the power factor is 1; at any other point the power factor is less than 1.

Unity power factor When voltage and current is perfectly in phase, the power factor is 1.

Azimuth Direction on a compass. Standard compass readings give us 0 azimuth for north, 90 degrees azimuth for east, 180 azimuth for south and 270 azimuth for west. In some solar resources there is a different way of reading azimuth which will find south as zero degrees; however, most people consider south to be 180 azimuth.

Magnetic declination A magnetic compass points toward the magnetic north pole, which is just north of the middle of Canada. The east coast has negative magnetic declination to the west and is corrected by rotating clockwise; the west coast of the USA has positive magnetic declination to the east and is corrected by rotating counter-clockwise. Also, for negative declination we can subtract from magnetic azimuth to get true azimuth; for positive declination we can add to magnetic azimuth to get true azimuth. For instance, in a location in California where there is 15 degrees of magnetic declination to the east, if the needle of the compass says 180 degrees, we add 15 and it is really 195 degrees, which would be a great direction to face your PV!

kWh/kWp/yr kWh produced per kW of PV installed in a year. Often it is useful to know how many kWh a kW will make in a year when sizing PV systems or determining if a system is economically feasible. The p from kWp stands for peak sun or 1000W/square meter STC conditions. Typical kWh/kWp/yr for the USA ranges from 1200kWh/kWp/yr to 1600kWh/kWp/yr with good solar access. With shading and north-facing arrays, it can be much worse. If we have a 2kW PV system in a location with 1500kWh/kWp/yr, then the expected annual production would be 2kW × 1500kWh/kWp/yr = 3000kWh per year.

Gross income All income before expenses and deductions.

Net income Income minus expenses.

Tax deduction A business expense that can be deducted from the gross income. If a person is in a 30% tax bracket, then the value will be about 30% of the tax deduction, realized through tax savings. This is not as good as a tax credit.

Depreciation As the value of a business asset decreases, then the asset can be depreciated for its loss in value. This loss can be a tax deduction.

Tax credit Whatever amount the tax credit is, it can be used at full value for paying taxes. A tax credit is better than a deduction or depreciation.

ITC (investment tax credit) The ITC has been in place for solar PV and has been a political incentive for solar over the years. The ITC is based on a percentage of the cost of the project. Some solar companies that own systems on other people's properties have been in trouble with the Internal Revenue Service for inflating the cost of their systems in order to get a greater tax credit. The credit will go to the owner of the system and there are many creative ways that have been developed to allow the tax credits to go to different entities. Go to www.dsireusa.org for the latest updates on the US ITC and other solar incentives. Historically, the solar ITC has been renewed at the last minute.

PTC (performance tax credit) A tax credit based on the performance of a renewable energy system. The wind industry has benefited from a PTC.

PBI (performance based incentive) A PBI is an incentive, such as a PTC, which will give a renewable energy system owner an incentive based on production. A PBI gives a solar system owner a better incentive to boost performance than an investment-based incentive. Some rebates are based on performance.

MACRS (modified accelerated cost recovery system) This is a form of accelerated depreciation. Cost recovery for depreciation is five years for solar. After the five-year term, the tax equity is used and the project can be bought and sold by those not needing tax equity, such as the building owner.

Supply-side connection Connecting the PV inverter between the meter and the main service disconnect. You can connect as much as the service can handle on the supply side. Often incorrectly called a "line side tap." A supply-side connection is very much like a separate utility service and can be connected using Article 230 Services of the NEC.

Load-side connection Connecting PV on the load side of the main service disconnect. Examples of load-side connections are the 120% rule, feeder connections, feeder taps and subpanel connections. Anything that is sharing overcurrent protection with other loads is a load-side connection.

120% rule Busbar $\times$ 1.2 $\geq$ Main breaker + (inverter current $\times$ 1.25).

The 120% rule refers to load-side interconnections where the solar breaker must be placed on the opposite side of the busbar from the main breaker. The 2011 and earlier Code uses inverter breaker in place of (inverter current $\times$ 1.25). The 120% rule should appear many times on the NABCEP Tech Sales Exam. If we do not exceed the busbar with our calculations, then we do not have to apply the 120% rule and can place our inverter breaker anywhere on the busbar.

Automatic transfer switch A switch that will automatically switch to an alternative power source when the grid is down. It is recommended to connect utility interactive inverters to the supply side of a generator automatic transfer switch. If a typical generator is running and connected to an interactive inverter, the inverter will not have a clean enough sine wave to turn on and the dirty sine wave might not be good for the inverter.

Qualified person One who has received training in and has demonstrated skills and knowledge in the construction and operation of electrical equipment and installations, and the hazards involved.

PPE (personal protective equipment) Fall protection, gloves, goggles and other equipment used to protect workers from injury.

Inverter breaker sizing Inverter current $\times$ 1.25 and round-up to the next common breaker size.

Common service voltages Residential 120/240V; small commercial 120/208V; large commercial 277/480V or 480V delta.

Interconnection agreement Agreement with a utility to connect the system. Usually takes place after the system is installed and the building permit is signed off by an inspector.

IRR (internal rate of return) The rate of return of a project without considering interest rates or inflation.

SREC (solar renewable energy certificate) or REC For each MWh produced, a solar system owner will earn an SREC, which they usually sell to a utility that is mandated to have solar in their energy portfolio. The REC system is common on the east coast of the USA.

Net-metering An arrangement with the utility in which the customer gets credit for energy exported when they are using less electricity than they are making. According to the Solar Energy Industries Association (SEIA), on average only 20–40% of energy produced ever gets exported. In many places with annual net-metering, you can get credit for energy made in the summer and use the energy in the winter. Some places do not have net-metering.

DSIRE (Database of State Incentives for Renewables and Efficiency) This is found at www.dsireusa.org. This is also where you can find out about various US state incentives for renewable energy. It is a good place to study online.

IREC (Interstate Renewable Energy Council) A non-profit organization that credentials training programs and instructors, and generates information about the clean-energy industry.

PACE (Property Accessed Clean Energy) Financing from local governments in which the money is paid back through a special assessment of property taxes. There is a lien on the property and if there is a foreclosure or new owner, they will still have to pay for the solar. It is organized much like municipal improvements, such as an increase in property taxes for a new sewer system in the neighborhood.

Compounding interest

Chapter 2

Many people think that if they invested $100 and made 10% interest per year, that they would have made $100 at the end of ten years. In reality, it is much better than this, unless you are the one borrowing the money.

We can see from the example in Table 2.1 that due to a compounding interest rate of 10%, in just over eight years our money has doubled and our interest has also doubled from what it was in the beginning.

Table 2.1 $100 with a 10% interest rate for ten years

YEAR	MONEY AT BEGINNING OF YEAR	10% INTEREST (BEGINNING OF YEAR MONEY × 0.1)	MONEY AT END OF YEAR (ADD INTEREST TO BEGINNING OF YEAR MONEY)
1	$100	$10	$110
2	$110	$11	$121
3	$121	$12.10	$133.10
4	$133.10	$13.31	$146.41
5	$146.41	$14.64	$161.05
6	$161.05	$16.11	$177.16
7	$177.16	$17.72	$194.87
8	$194.87	$19.49	$214.36
9	$214.36	$21.44	$235.79
10	$235.79	$23.58	$259.37

There is a shortcut here; rather than making your own table, we can just punch a few keys on the calculator. In the NABCEP Technical Sales Exam, you will be given a Casio fx260 calculator, which you can find for less than $10 on the internet, so I recommend you get one. It is also a solar-powered calculator, so it will work off-grid!

To get a 10% increase, what we do is multiply by 1.1. If we are going to get a 10% increase ten different times, then what we do is multiply by 1.1 ten times.

$$1.1 \times 1.1 \times 1.1 \times 1.1 \times 1.1 \times 1.1 \times 1.1 \times 1.1 \times 1.1 \times 1.1 = 2.59$$

This gives us 2.59 times what we invested after ten years; since we invested $100 then $2.59 \times \$100 = \$259$

An easier way of accomplishing this is with your calculator buttons.

Look for the X^y button, which is the trick.

- Enter: 1.1
- Enter: X^y
- Enter: 10
- Enter: =

This should give you 2.59; now all you have to do is multiply 2.59 by your original investment, which was $100:

$$2.59 \times \$100 = \$259$$

> To turn a percent into a decimal, divide by 100. It is silly that someone invented the percentage system. We should have just used the decimal system in the first place, so we don't waste ink with extra zeros and moving decimals two places all of the time.

Calculating interest can also be done on a sales proposal when we are comparing how much money someone could have made investing it, rather than paying the utility. There are a lot of ways this works.

When we have system losses, the rate can also compound, which also can be a greater benefit than the alternative. If we lost 50% of $100 for two years, on year 1 we would have $50 and on year 2 we would have 50% of $50, which is

$25. $25 is much better than zero and some people would incorrectly think that losing 50% two years in a row would leave nothing.

If we have module degradation, we will have slight losses over the years. How much those losses are is debatable, but for the next exercise, let us conservatively estimate that our losses due to degradation are going to be 1% per year. In reality it may be closer to 0.5% per year, but that depends upon the module selected.

If we have a module that has a degradation rate of 1% per year for 25 years, then we can calculate the loss of performance by using this formula, but instead of using 1.1 for a 10% increase we need to use a number for a 1% decrease.

If we lose 1%, then we keep 99%. Usually we are more interested in what we keep than what we lose. That is also a good way to look at life.

If we have 99% for 25 years, then we can calculate that by

$$0.99 \times 0.99 \times 0.99 \times 0.99 \times 0.99 \times 0.99 \times 0.99 \times 0.99 \times$$
$$0.99 \times 0.99 \times 0.99 \times 0.99 \times 0.99 \times 0.99 \times 0.99 \times 0.99 \times 0.99 \times$$
$$0.99 \times 0.99 \times 0.99 \times 0.99 \times 0.99 \times 0.99 \times 0.99 \times 0.99 = 0.78$$

This means that if we lose 1% per year for 25 years, we end up losing about 22%, rather than the worse scenario of losing 25% that many normal people would have assumed (people who have an aversion to addition and subtraction).

We can more easily do this by entering into our calculator:

- Enter: .99
- Enter: X^y
- Enter: 25
- Enter: =

And then you have 0.78, which is the decimal equivalent of keeping 78% or losing 22%.

It is often easier to make the point with more extreme numbers.

If you had a dollar and an interest rate of 100% per year (which is called doubling), then in one year you would have $2; in three years you would have $4; and in four years you would have $8:

- 100% more than $1 is $2
- 100% more than $2 is $4
- 100% more than $4 is $8

On the calculator, you add 1 + 1 to give you the formula for a 100% increase (as easy as 1 + 1 = 2).

- Enter: 2
- Enter: X^y
- Enter: 3
- Enter: =

The answer here is 8, since 2^3 is 8.

100% growth is common in the solar industry. Some years we have 100% job growth, 100% PV installed growth and, once you finish this book, it is predicted that you will have 100% income growth, because by using your brain you can make yourself expensive!

Let's see how much we *keep* in this next example:

- If we are losing 25% per year for three years, then how much do we keep?
- Losing 25% means keeping 75% (1 – 0.25 = 0.75)
- We can accomplish this by calculating:

$$0.75 \times 0.75 \times 0.75 = 0.42 \text{ or } 42\%$$

- If you lose 25% per year for three years, then you lose 58% altogether.

The following example relates to a doubling rate, known as 100% annual growth:

Now we will get ourselves excited about the exponential growth of the solar industry. Let's say that the amount of PV that exists in the world doubles every two years, which is not unusual. For this one our time period is two years instead of one year.

If the amount of PV doubles every two years and we estimate that 1% of the energy in the world is currently made by PV, then how much energy will be made by PV in 20 years?

This time we are going to take 1% and see how much we can increase 1% by doubling 1% ten times, since we double ten times in 20 years, with 10 two-year periods.

$$1 + 1 = 2 \text{ for a 100\% increase}$$

We can now multiply $2 \times 2 \times 2 \times 2 \times 2 \times 2 \times 2 \times 2 \times 2 \times 2 = 1024$

Or we can use the shortcut: $2^{10} = 1024$

So we are going to have 1024 times what we started with and since we started with 1%, then in 20 years (ten doublings) we are going to have 1024% of the energy we use today made from PV. That is going from 1 to over 1000 with ten doublings – and that is called exponential growth!

If you are reading this in 20 years, please let me know what happened. There is a good chance that when 100% of energy production is reached within seven doublings ($2^7 = 128$), that there will be less of a demand for energy, but at least you can see the trajectory we are following and know that you are in the right profession!

Now we will show you the textbook formula and terms that are used for these calculations.

$$PV \times (1 + r)^n = FV$$

Now to confuse you just a little bit, in this case PV means something other than photovoltaic.

- **PV = present value** or photovoltaic
- r = rate (interest rate in decimal form)
- n = number of periods (often this will be a year-long period)
- FV = future value

We can also play with this formula and solve for the present value if we have the future value.

If :

$$PV \times (1 + r)^n = FV$$

Then:

$$PV = FV/(1 + r)^n$$

Let us use these equations a few times to get ready for real life.

Would you rather have $1000 today or $2000 in ten years if you could make 8% interest off of your money?

Let's check it with the first formula and solve for FV and compare the calculated FV of the $1000 making 8% to the $2000 option:

- PV = present value = $1000
- r = rate = 8% – 0.08 in decimal form
- n = period = ten years
- FV = future value (is it greater than $2000?)

Here is the formula:

$$PV \times (1 + r)^n = FV$$
$$\$1000 \times (1 + 0.08)^{10} = FV$$
$$\$1000 \times 1.08^{10} = FV$$
$$\$1000 \times 2.15892 = FV = \$2158.92$$

$2158.92 is the future value of the $1000 investment, so we can make an extra $159 if we invested in this case and would be better off waiting ten years.

Let us do another calculation and see what we would have to invest in order to have $2000 ten years from now if we had an 8% interest rate.

- FV = $2000
- r = rate = 8% = 0.08 in decimal form
- n = period = ten years
- PV = unknown that we will invest

$$PV \times (1 + r)^n = FV$$
$$PV = FV/(1 + r)^n$$
$$PV = FV/(1 + 0.08)^{10}$$
$$PV = FV/(1.08)^{10}$$
$$PV = FV/2.1589$$
$$PV = \$2000/2.1589$$
$$PV = \$926$$

So $2000 in ten years with an 8% interest rate is worth $926 today.

These calculations do not take inflation into consideration, so it might be more fun to just spend your $1000 now.

Here, we do the inflation calculation, much like the compounding interest calculation.

Let's say that we want to see how much $2158.92 ten years from now is worth today due to an inflation rate of 3%.

Since it is worth less, we will be losing 3% per year on our money.

If we want to use our same formula.

- PV = unknown
- FV = \$2158.92
- n = ten years
- r = 3% = 0.03

$$PV \times (1 + r)^n = FV$$
$$PV = FV/(1 + r)^n$$
$$PV = FV/(1 + 0.03)^{10}$$
$$PV = FV/1.03^{10}$$
$$PV = FV/1.3439$$
$$PV = FV/1.3439$$
$$PV = \$2158.92/1.3439$$
$$PV = \$1606$$

Still, we can see that we are \$606 better off investing the money if we take inflation into consideration.

You can easily find an inflation and interest calculator online to practice. There are also *future inflation calculators* that you can use to double check your work. In reality, inflation is different every year and the best we can do is estimate future inflation.

One of the most common mistakes people make with PV math is adding and subtracting when they should be multiplying and dividing. This can happen on the busbar and in the financial analysis.

We can work this with a PV system degradation rate of 1% per year and compare it with a PV system degradation of 0.5% per year. This will allow us to see how much energy we will be making in year 10 in both scenarios if we start out making 1MWh annually on year 1.

I am going to work this problem out without writing down a formula, which some solar people find easier, since they are used to using derating factors from PVWATTS and the NEC.

To determine the losses at year 10, with a 1% per year loss, we can multiply the derating factor to be:

$$0.99^{10} = 0.904$$

Then we multiply our year 10 derating factor by 1MWh or 1000kWh (which is the same).

$1000kWh \times 0.904 = 904kWh$ per year with 1% per year degradation

Now we will determine what the year 10 production would be with 0.5% per year system losses.

If we lose 0.5% per year, then we keep 99.5% per year.

Mathematically, working with decimals, that is:

$$1 - 0.005 = 0.995$$

First we find our new derating factor:

$$0.995^{10} = 0.951$$

Then we multiply our year 10 derating factor by 1MWh or 1000kWh (same).

$1000kWh \times 0.951 = 951kWh$ in year 10 with 0.5% annual degradation

Common sense will always tell us that our number is going to increase or decrease. With derating factors, you will multiply or divide to get your answer. Pay attention to the obvious and do not get lost in the processes. Compounding interest is just taking multiple derating factors for each year, which are all the same, and multiplying them together with exponents.

Final conception:

$$1.1^{20} = 6.7$$

This means that for 10% interest in 20 years, which is the length of your average feed-in tariff (FIT), you will make 6.7 times your money. Now you know why the German bankers were so excited about the German FIT program!

Another reason that people like compounding interest is that it will pay for their retirement. Why not get a PV system to pay for your retirement or at least the electricity that you will use in your retirement?

CHAPTER 2 PRACTICE QUESTIONS

Here are ten practice problems that you can work out yourself. The answers will follow.

1. $100 invested at an interest rate of 5.3% would be worth how much in 22 years?

2. $1000 invested at an interest rate of 5% per year would be worth how much in 14 years?

3.   With a PV degradation rate of 1% per year, what would be annual production in 12 years if the system originally were making energy at a rate of 1500kWh/kWp/yr?

4.   With an inflation rate of 2.5%, what is a dollar worth five years from now?

5.   At an interest rate of 7%, how much will $1215 be worth in 17 years?

6.   If you were going to gain 15% per year on your investment of $103,000, how much money would you have after 33 years?

7.   If you were going to lose 15% per year and you started with $103,000, how much money would you have after 33 years?

8.   With an inflation rate of 3% per year, how much is $100 worth in six years?

9.   With a PV system degradation rate of 0.75% per year, if a system made 1MW when it was installed during peak sun conditions, what would the system produce during peak sun conditions 11 years later?

10. If you invested a penny for your distant offspring, how much would it be worth 1000 years from now with an interest rate of 10% per year?

DETAILED ANSWERS TO CHAPTER 2 QUESTIONS

1. $100 invested at an interest rate of 5.3% would be worth how much in 22 years?

We need to solve for future value.

- Present value = $100
- $r = 5.3\% = 0.053$
- $n = 22$ years

$$PV \times (1 + r)^n = FV$$
$$\$100 \times (1 + 0.053)^{22} = FV$$
$$\$100 \times (1.053)^{22} = FV$$
$$\$100 \times 3.11 = FV$$
$$\$311 = FV$$

$100 at an interest rate of 5.3% in 22 years will be worth $311.

2. $1000 invested at an interest rate of 5% per year would be worth how much in 14 years?

We need to solve for future value.

- Present value = $1000
- $r = 5\% = 0.05$
- $n = 14$ years

$$PV \times (1 + r)^n = FV$$
$$\$1000 \times (1 + 0.05)^{14} = FV$$
$$\$1000 \times (1.05)^{14} = FV$$
$$\$1000 \times 1.98 = FV$$
$$\$1980 = FV$$

$1000 at an interest rate of 5% in 14 years will be worth $1980.

3. With a PV degradation rate of 1% per year, what would be annual production in 12 years if the system originally were making energy at a rate of 1500kWh/kWp/yr?

We need to solve for future value.

- Present value = 1500kWh/kWp/yr
- $r = -1\% = -0.01$
- $n = 12$ years

$$PV \times (1 + r)^n = FV$$
$$1500\text{kWh/kWp/yr} \times (1 - 0.01)^{12} = FV$$
$$1500\text{kWh/kWp/yr} \times (0.99)^{12} = FV$$

$$1500\text{kWh/kWp/yr} \times 0.886 = \text{FV}$$
$$1329\text{kWh/kWp/yr} = \text{FV}$$

A PV system that makes 1500kWh/kWp/yr when new will make about 1329kWh/kWp/yr after 12 years with a system degradation rate of 1% per year.

4. With an inflation rate of 2.5%, what is a dollar worth five years from now?

We need to solve for future value.

- Present value = $1
- $r = -2.5\% = -0.025$
- n = five years

$$\text{PV} \times (1 + r)^n = \text{FV}$$
$$\$1 \times (1 - 0.025)^5 = \text{FV}$$
$$\$1 \times (0.975)^5 = \text{FV}$$
$$\$1 \times 0.88 = \text{FV}$$
$$\$0.88 = \text{FV}$$

$1 with an inflation rate of 2.5% in five years will be worth $0.88.

5. At an interest rate of 7%, how much will $1215 be worth in 17 years?

We need to solve for future value.

- Present value = $1215
- $r = 7\% = 0.07$
- n = 17 years

$$\text{PV} \times (1 + r)^n = \text{FV}$$
$$\$1215 \times (1 + 0.07)^{17} = \text{FV}$$
$$\$1215 \times (1.07)^{17} = \text{FV}$$
$$\$1215 \times 3.16 = \text{FV}$$
$$\$3839 = \text{FV}$$

$1215 at an interest rate of 7% in 17 years will be worth $3839.

6. If you were going to gain 15% per year on your investment of $103,000, then how much money would you have after 33 years?

We need to solve for future value.

- Present value = $103,000
- $r = 15\% = 0.15$
- $n = 33$ years

$$PV \times (1 + r)^n = FV$$
$$\$103,000 \times (1 + 0.15)^{33} = FV$$
$$\$103,000 \times (1.15)^{33} = FV$$
$$\$103,000 \times 101 = FV$$
$$\$10.4M = FV$$

$103,000 at an interest rate of 15% in 33 years will be worth $10.4 million dollars!

7. If you were going to lose 15% per year and you started with $103,000, then how much money would you have after 33 years?

We need to solve for future value.

- Present value = $103,000
- $r = -15\% = -0.15$
- $n = 33$ years

$$PV \times (1 + r)^n = FV$$
$$\$103,000 \times (1 - 0.15)^{33} = FV$$
$$\$103,000 \times (0.85)^{33} = FV$$
$$\$103,000 \times 0.00469 = FV$$
$$\$483 = FV$$

$103,000 at an interest rate of –15% in 33 years will be worth $483.

8. With an inflation rate of 3% per year, how much is $100 worth in six years?

We need to solve for future value.

- Present value = $100
- $r = -3\% = -0.03$
- $n = 6$ years

$$PV \times (1 + r)^n = FV$$
$$\$100 \times (1 - 0.03)^6 = FV$$
$$\$100 \times (0.97)^6 = FV$$
$$\$100 \times 0.833 = FV$$
$$\$83 = FV$$

$100 with an inflation rate of 3% in six years will be worth $83.

9. With a PV system degradation rate of 0.75% per year, if a system made 1MW when it was installed during peak sun conditions, what would the system produce during peak sun conditions 11 years later?

We need to solve for future value.

- Present value = 1MW
- $r = -0.75\% = -0.0075$
- $n = 11$ years

$$PV \times (1 + r)^n = FV$$
$$1MW \times (1 - 0.0075)^{11} = FV$$
$$1MW \times (0.9925)^{11} = FV$$
$$1MW \times 0.9205 = FV$$
$$920.5kW = FV$$

A PV system that makes 1MW during peak sun conditions, with a degradation rate of 0.75% per year, can be expected to make 920.5kW 11 years later during peak sun conditions.

10. If you invested a penny for your distant offspring, how much would it be worth 1000 years from now with an interest rate of 10% per year?

We need to solve for future value.

- Present value = $0.01
- $r = 10\% = 0.1$
- $n = 1000$ years

$$PV \times (1 + r)^n = FV$$
$$\$0.01 \times (1 + 0.1)^{1000} = FV$$
$$\$0.01 \times (1.1)^{1000} = FV$$
$$\$0.01 \times (2.47 \times 10^{41}) = FV$$
$$\$2.47 \times 10^{39} = \$2,470,000,000,000,000,000,000,000,000,000,000,000,000$$

This is way more than all of the money in the world, which is about 2.5 duodecillion dollars. Remember to save your pennies.

Electricity

You can sell a car without understanding the difference between power and energy, but if you are selling solar, you are selling energy and it is good for your reputation to know what you are selling, especially if you get what solar salespeople will tell you is the most difficult customer: someone who is literate in electricity. Solar salespeople fear customers that are engineers. Pay attention and you can be one of the few solar salespeople that can sell solar to engineers.

What is electricity?

Electricity is a vague term used to mean power, energy, voltage or any movement of electrons. If you do not understand the difference between power, energy, voltage and current, then play it safe and call it electricity (this will not work on an exam).

One of the first things you will ask a prospective customer is to take a look at 12 months of utility bills to determine their energy usage. It always amazes me how many people in the industry do not know the units of energy.

Energy can be stored; power cannot be stored. Energy is a quantity; power is a rate. You can still sell power and in some cases the utility will have demand charges for large customers for power, but when you go home and write a check to the utility, unless you live in a strange place, you are not paying for power, you are *not* paying for kilowatts (kW), you are paying for energy, you are paying for kilowatt-hours (kWh)!

When you ask your customer how much energy they use, they might say they used 1000 kilowatts last month, but you should know that they meant they used 1000 kilowatt-hours. I will usually correct a customer one or two times, but after that they need to sign up for a class if they want to be educated. Either way, I will still save them money, which they can earn interest on, by selling them a rooftop power plant.

Notice how I just said power plant? When we talk about how large the power plant is, we can say that in power terms. It does not tell us how much money we save unless we know how long the power plant is on and how hard it is working. Just like when you buy a brand new Tesla Model X next year and they tell you that it can go 150mph, you will not usually go that fast. More often than 150mph, you will be driving at half that speed. This is much how the power plant on your roof works.

Some power plants can make a constant amount of power 24 hours per day. PV plants will make a variable amount of power during the day and no power during the night. In fact, a 5kW PV system will never put out 5kW, because there are always losses and because we rate PV based on test conditions that are not normal environmental conditions that our array will operate under. It is common that you will be looking at the power coming out of your inverter in the middle of the day and will see a 5kW dc system producing 3 ac kW.

One thing that gets people mixed up about kWh is that they are so used to driving and seeing those signs that say mph and think that kWh also has to be a rate. Miles per hour is a rate and could also be written as miles/hour. If we multiplied miles/hour by hours we then get miles, since the hours cancel each other out:

$$\text{miles/hour} \times \text{hours} = \text{miles}$$

In the case above, miles/hour was the rate and miles were the quantity.

In the case below, we are going to do something different and multiply the rate of electricity, which is power (given in kW), by time to get the quantity of electricity, which is energy (given in kWh).

$$\text{power} \times \text{time} = \text{energy}$$
$$\text{kW} \times \text{hr} = \text{kWh}$$

Let us sum it up:

$$\text{rate} \times \text{time} = \text{quantity}$$
$$\text{miles/hour} \times \text{hour} = \text{miles}$$
$$\text{speed} \times \text{time} = \text{distance}$$

$$kW \times hours = kWh$$
$$power \times time = energy$$

The next lesson:

$$power = voltage \times current$$

From looking at this equation and the last one, we can infer that since

$$power \times time = energy$$

and

$$power = voltage \times current$$

then

$$voltage \times current \times time = energy$$

CURRENT

Current is represented by the symbol I and is measured in amps or amperes.

Current is a simple concept and relates well to the current of a river. Current is the flow of electrons. Many exams will ask for the hydraulic analogy for current, which is flow.

The amount of flow is impossible to count unless you are as fast as the computer chip that is writing this book.

1 amp = 6.2×10^{18} electrons per second passing by.

"Amp" is the shortened form for ampere, so an amp is an ampere and is a measurement of current. Current is symbolized by the letter I, which comes from the word intensity – but I like to think like a solar guy and figure that, for us, I comes from irradiance (the brightness of light).

Current is important, since current is what will heat up equipment and conductors (wires). If we have a lot of current on a wire, it will cause heat, loss of power and voltage drop.

VOLTAGE

Voltage is like pressure. Voltage is electrons wanting to go where there are no electrons; the more voltage there is, the more they want to get moving.

If there is enough voltage, the electrons can travel through things that we usually do not want them to go through, like air, PV module glass, insulation on wires, from wire to wire or, worst of all, through you.

Since voltage × current = power, if we have more voltage, then we have less current for the same amount of power. With less current, we have less power loss and less heating up of the wire, which is great for transmission over long distances, but not so good for keeping sparks from ruining all of the electrical equipment in your house.

Often, waterfall analogies are used to represent voltage and current (Figures 3.1 and 3.2).

Figure 3.1 High-voltage, low-current waterfall analogy: Angel Falls. © Flickr/ David Kjelkerud/https:// creativecommons.org/ licenses/by/2.0.

Figure 3.2 High-current, lower-voltage waterfall analogy: Niagara Falls. Source: Frederic Edwin Church 1857 (https://upload.wikimedia.org/wikipedia/commons/c/c4/Frederic_Edwin_Church_-_Niagara_Falls_-_WGA04867.jpg).

AMP HOURS

When we are buying or sizing batteries, there is a concept that is not often used in other places that refers to **battery capacity**, which is **measured in amp hours (Ah)**. It is sort of like energy, but not accounting for voltage. We could measure loads with amp hours, but nobody bothers, except classic battery people.

An amp for an hour is an amp hour. If we have a 12V battery bank giving an amp for an hour, that would be the same amount of amp hours as a 48V battery bank giving an amp for an hour. The 48V battery bank, however, would give us four times as much energy, since an amp for an hour at 48V is four times more energy than an amp for an hour at 12V.

If you are battery shopping next year for a customer that just purchased a battery backup system and you find a deal on 6V batteries, how will you know if it is better or worse than the deal on 12V batteries? If each battery has 100Ah are they equals? It also depends on the price and quality of the batteries.

Here are some related equations to ponder and know:

$$\text{watts} \times \text{hours} = \text{watt hours}$$
$$\text{volts} \times \text{amps} = \text{watts}$$
$$\text{volts} \times \text{amps} \times \text{hours} = \text{watt hours}$$

$$\text{amps} \times \text{hours} = \text{amp hours}$$
$$\text{amp hours} \times \text{volts} = \text{watt hours}$$

When connecting batteries, or solar modules, in series, the voltage of each is added and the current remains constant. When a connection is parallel, the current is added and the voltage remains constant. For instance, if I had two 12V 100Ah batteries and I hooked them up in series, I would then have a 24V 100Ah battery bank. If I had both of those batteries connected in parallel, then I would have a 12V 200Ah battery bank.

Either way, it still consists of the same batteries and would still be the same amount of kWh, however you calculate it.

Here are three different ways of calculating the energy:

- 2 batteries $\times$ 12V $\times$ 100Ah = 2400Wh = 2.4kWh
- 24V $\times$ 100Ah battery bank = 2.4kWh
- 12V $\times$ 200Ah battery bank = 2.4kWh

For these different systems, it would usually be more efficient to have the batteries connected in series, since higher voltage means lower current and less losses. The benefit of lower voltage is that there is a lower likelihood of a shock through your body, since the voltage would have trouble getting through the resistance of your body.

Figure 3.3 Batteries with series connections: 24V 100Ah. Image by Sean White.

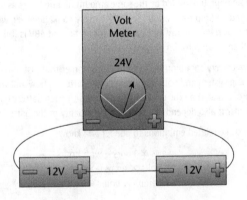

Figure 3.4 Batteries with parallel connections: 12V 200Ah. Image by Sean White.

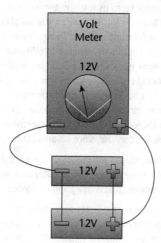

To summarize:

Current is the flow of electrons. Too much flow will cause a wire to heat up, since all of those electrons cannot fit in a small wire. Current is measured in amperes (A or amps for short) and is symbolized by the letter I for intensity; current is increased by irradiance (brightness).

Voltage is "electric pressure"; if the electrons have too much pressure, they will be inclined to jump the gap and cause an arc, which is why we do not have high-voltage wires running to our toaster. Voltage is measured in volts and sometimes we see scientists using the letter E to represent voltage, which comes from the physics term electromotive force (EMF).

Power is the instantaneous rate of use of electricity or mechanical processes. If we measure our car's power or a pump, we often use the term *horsepower*; 746W = 1 horsepower. Solar power is usually measured in watts (W) or kW; we can use the symbol P to represent power. An Olympic power lifter might only be able to lift the heavy weights once; if they can lift the weights many times, then we can call them an energy lifter.

Energy is power times time, also known as work. Energy is so important that wars are fought for it. Energy keeps us warm in the winter and cool in the summer. Energy comes in all forms. In this book we mostly talk about electricity and kWh, but there is energy in food and gasoline. There are about 33kWh in a gallon of gas. If we are talking about food, there are 1.16Wh in a Cal. This still does not mean that we can drink gasoline, but we can use gasoline to grow food with lights, but there would be too many losses along the way. In this book we will refer to energy as watt hours (Wh) or more commonly as kWh, which is what we pay for with our electric bill and what we want to know about our prospective customers. A good solar salesperson will never ask their typical prospective customer how many kW they use, since we are in the energy business.

In the solar sales business, you will probably have to deal with customer resistance on a daily basis; however, you will probably deal with electrical resistance a lot less than the engineer that is designing the system.

If you are selling a system that is far away from the meter you are feeding, you will have to pay more for wire, since the wire is longer. The wire also has to be thicker, due to losses from resistance. We call this effect voltage drop. There is an equation called Ohm's law which describes the relationship between resistance, voltage and current.

OHM'S LAW

$$voltage = current \times resistance$$

Resistance is measured in Ohms; we can also look at it this way:

$$resistance = volts/amps$$

When we are transporting electrons over a long distance, we want as little resistance as possible. A thicker wire has a lower resistance and a longer wire has more resistance.

Let's look at voltage drop and see how this affects our bottom line:

$$Voltage\ drop = operating\ current \times resistance\ of\ wire$$
$$Vdrop = IR$$

We have a constant resistance within our wire run (the wire does not get thicker as we add current). Therefore, if we increase the current along the wire, such as with brighter sunlight, we will increase our voltage drop. Increasing current increases our losses and in order to make up for increased current we need to increase the size of our wire.

Voltage drop percentage is the volts that we have lost due to increasing current over a wire compared to the system voltage.

Voltage drop percentage = (voltage drop/system voltage) × 100%

Say we have a wire and that wire is thin; we are going to lose some volts over that wire when we try to push a lot of electrons through it.

$$Vdrop = IR$$
$$I = 10A$$
$$R = 1 \text{ ohm}$$

Solve for Vdrop:

$$Vdrop = 10A \times 1 \text{ ohm} = 10V$$
$$\text{System voltage} = 240V$$
$$\text{Voltage drop percentage} = (10V/240V) \times 100\%$$
$$\text{Voltage drop percentage} = 4.2\%$$

This means that on the side of the wire of the source, there are 4.2% more volts measured with our voltmeter than on the side of the wire where the power is going.

If we lose 4.2% of our voltage, we will lose 4.2% of our power, 4.2% of our energy production and 4.2% of the value of our system.

It is important for a solar salesperson to know that a long wire run with an undersized wire can lead to some serious cuts in production.

Another interesting thing about voltage drop is that the voltage is always higher at the source of the power. That means that if we have PV feeding an inverter, the voltage will be higher at the PV, in order to get the energy to the inverter. Also, even more interesting, the voltage at the ac side of the inverter will have to be higher than the voltage at the interconnection of the PV system. That means that if we have a 240V grid and are losing 10V, then we will have 250V at the inverter. If the voltage gets too high at the ac side of the inverter, the inverter will shut off. This is why we have to make sure we do not have too much voltage drop on the ac side of the inverter. Sometimes the voltage drop is on the utilities side of the meter, which is good, since we do not have to pay for the losses, but can be bad if the voltage rises so much that the inverter shuts off.

One thing about current on a PV system is that it is variable with regard to the brightness of the light that hits the PV cells. As it gets brighter, the voltage drop will increase. Beware of lost (and rising) volts on bright days!

OVERVIEW

$$V \times I = P$$
voltage $\times$ current = power

$$V \times A = W$$
volts $\times$ amps = watts

$$W \times hr = Wh$$
watts $\times$ hours = watt hours

$$1000W = kW$$
$$kW \times hr = kWh$$

$$A \times hr = Ah$$
amps $\times$ hours = amp hours

$$A \times hr \times V = Wh$$
amps $\times$ hours $\times$ volts = watt hours

$$Ah \times V = Wh$$
amp hours $\times$ volts = watt hours

$$V = I \times R$$
voltage = current $\times$ resistance

$$V = A \times Ohms$$
volts = amps $\times$ Ohms

$$1000 = k$$
$$1000k = M$$

energy = kWh = $

knowledge = power $\times$ time = energy = kWh = $

Photovoltaics

Chapter 4

Photo = light.

Voltaic = V.

Photovoltaic means making voltage from light.

Photons are pieces of electromagnetic energy that travel through the air. Scientists have changed their minds over time, calling light a particle, then a wave and now a particle-wave.

Light is quantized and is in packets. You can have one photon, you can have two photons, but you cannot have 1.5 photons. This means that light comes in pieces and you cannot break apart the pieces.

There are many spiritual people talking about good energy and quantum theory that are talking about another language than physics. In physics, energy is power × time; quantum theory just means that when you get small enough, you cannot break things down any more. There is no such thing as 1.5 photons. Quantum deep!

Figure 4.1 shows different ends of the solar electromagnetic spectrum; we can see that visible light lies between the higher frequency/shorter wavelength and more dangerous UV light at one end, and the safer lower energy/longer wavelength infrared light at the other end.

Long wavelength photons can be radiowaves, such as from the cell phones we hold next to our brains. The longer the wavelength, the lower the energy. High-frequency wavelengths can be X-rays and gamma rays, which are hazardous to your health. We can see that the peak of the energy comes in the visible

37

Figure 4.1 Solar radiation spectrum. Source: Robert A. Rohde (https://upload. wikimedia.org/wikipedia/commons/4/4c/Solar_Spectrum.png).

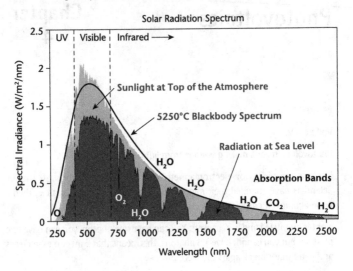

part of the spectrum, which is probably why our eyes evolved to see "visible light."

PV tends to work better in the visible spectrum of light. Crystalline PV works a bit better on the red side of the visible spectrum (closer to infrared) and that is why crystalline PV is often bluish in color. Since crystalline PV is not good at turning blue light into electricity, we might as well put on a coating that will absorb the good red light and get rid of the unusable blue light, which would otherwise create **heat** at the solar module, because heat decreases voltage, power, energy and income.

Thin-film PV has a reputation for doing better with blue light than does crystalline PV, and can absorb the blue side of the spectrum better, which is more prevalent on cloudy and overcast days. They say a MW of thin film might take up more space and resources, but it will also outperform most MWs of crystalline PV when it is not sunny. Additionally, thin-film does not react as much to losses of voltage when it heats up.

Another interesting thing we can see in Figure 4.1 is that there are two shades of the peak; the upper peak is sunlight outside of Earth's atmosphere and the lower peak is sunshine after being filtered by Earth's atmosphere. We can see that different molecules in the atmosphere filter out different parts of the spectrum. If we could get rid of the atmosphere, our solar would work better, but we would not work at all (breathe). PV would work better on the moon than on Earth, since there would be no air to get in the way of the photons.

When the sun gets lower in the sky, there will be more air filtering the light between the sun and us. This filtering effect will increase and that is why at sunrise and sunset you can look at or very close to the sun. Filtering is why, when the sun is lower, we are not as much affected by it. We call this filtering effect air mass or AM. There are test conditions under which PV is tested so that we can compare PV made in different places. All of the PV that you will work with is tested under the same standard test conditions (STCs).

STANDARD TEST CONDITIONS

* 1000W per square meter
* 25C (77F)
* 1.5 AM

Let's explain further.

1000W/SQUARE METER = SOLAR POWER = IRRADIANCE

If it were noon on a clear day, we would have about 1000W per square meter in many places (assuming that square meter was directly facing the sun). In some places we have more light and in other places we have less. One thousand is a very round number, so the people who came up with STC picked it. Usually, we have less solar power (irradiance) than 1000W/square meter, since it is not always noon on a clear day, but we have to account for more in order to be safe. More irradiance correlates with greater current (amps).

25C = CELL TEMPERATURE = 77F

When the people came up with 25C cell temperature test conditions for PV, it was not hot enough to truly represent what most of the solar modules are doing all day: sitting in the sun. It is a good energy-saving solution, since the warehouse where they are holding the PV and doing the testing is closer to 25C and the manufacturers do not have to put the modules in ovens to duplicate test conditions. It is also good for solar salespeople that the temperature is lower, because this way is makes our PV sound more powerful. If we tested at hotter temperatures, due to the loss of voltage the power would be less. On a summer day, our PV may be making about 15% less energy due to heat decreasing our voltage, maybe even less.

1.5AM (AIR MASS)

If the sun were straight over your head and you were standing at sea level, by definition you would have 1AM filtering the light between you and the sun. If you are not in the tropics (less than 23.5 degrees latitude, which is the tilt of the Earth), then you will never have the sun over your head. The sun can only be over your head in the tropics and even there it happens rarely (twice per year).

The STC inventors decided to pick 1.5AM for testing conditions since it was more reasonable. Another thing they had to do was standardize the spectrum of light, because besides the thickness of the air they would have to also standardize the quality of the air, which includes humidity, particles, smog, ozone and carbon dioxide. What they decided to do was measure the spectrum of light at Cape Canaveral, Florida, on equinox at noon and declare: "This is STC 1.5AM!" and henceforth STC was standardized.

Modules are tested at STC by a big box machine called a flash tester, which typically contains a bright xenon flashbulb, a light filter and leads to connect the module to, so that the parameters can be checked. The parameters are checked by running an IV curve test and recording the characteristics of the module.

IV CURVES

IV curves are run on solar cells, modules, PV source circuits and possibly arrays. An IV curve test on PV in the field can often tell the health of the circuit being tested. If there are shading problems, broken diodes or other malfunctioning parts of the module, an IV curve is often the best way to find the problem.

Figure 4.2 IV curve. Image by Sean White.

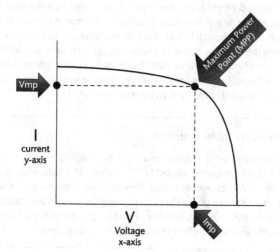

IV curves will determine the different combinations of voltage and current that can occur.

In the IV curve in Figure 4.2, we can see the horizontal x-axis represents voltage. On the right lower side of the curve, we see Voc, which stands for open circuit voltage. With electricity, "opening a circuit" means turning off. At Voc, our system is turned off, since we cannot pass current when there is no connection. We can also say that an open circuit represents infinite resistance. There is no way an electron can pass, there is just too much resistance through the air. An IV curve tester will essentially lower the resistance from infinite resistance to no resistance and measure all of the parameters in between.

The parameter with no resistance to current is called the short circuit current and is represented by Isc. At Isc, there is no voltage and therefore no power or energy. Isc is a direct short and we can represent this by taking the positive and the negative of a PV module and connecting them together. Since the conductors are connected together when they are shorted, then there is going to be no voltage (difference in electrical pressure) from the positive wire to the negative wire.

There are many places in between Voc and Isc where we can make power, energy and money, but there is one sweet spot where we can make the most. We call the sweet spot, where the product of voltage and current is the most power, the maximum power point (MPP). All grid-tied inverters have maximum power point tracking (MPPT). Many charge controllers operate at the MPP with MPPT. Many of the newer inverters have dual MPPT inputs and can track different parts of the array differently, which is convenient since they can work at different voltages. Only equipment that has PV as an input will have MPPT.

Non-MPPT charge controllers

Some older style or new, inexpensive charge controllers do not have MPPT. The charge controllers without MPPT are often called pulse width modulation (PWM) charge controllers. They can control the charge by quickly switching on and off the PV–battery connection. These charge controllers are about 20–30% less effective at producing power from the array, but are the logical, cost-effective choice for very small systems.

Shading on a PV module will take the smoothness out of an IV curve and make it look like the graph of a bad year on the stock market. Different inverters have different MPPT programs and will track differently in response to shading. If there is a shadow on a PV module, it will often look like there are different peaks on the curve where there are possibilities for MPPT.

BYPASS DIODES

Most PV modules have three bypass diodes which divide the modules into three sections lengthwise.

In Figure 4.3 we can see how solar cells are arranged into three groups in a typical 60-cell module. The groups are labeled A, B and C. If there is a shadow over one or all of the cells in a group, then the bypass diode will take the current past the shaded group of cells, keeping the current going through the rest of the groups of cells. When there is no current, the bypass diodes will not bypass, since there is nothing to bypass, which is why you can measure the open-circuit voltage of a

Figure 4.3 Solar cells are arranged in groups with bypass diodes.

shaded module and see that there is not a significant decrease in voltage. When the module is operating, however, perhaps at Vmp, then there will be a very significant decrease in voltage. PV gets close to full voltage in low light, but not much current.

If there is a shadow covering a row of cells on the long edge of the module, it will take out one-third of the module's operating voltage. If that same shadow was taking a row of cells along the short edge of the module, all three of the bypass diodes in the module would kick in and there would be no power contribution from the module.

> Bypass diodes will sacrifice voltage for current. Without current, there will be no power. Without voltage from a group of cells, there will still be some voltage and there will be some power.
>
> Bypass diodes are like taking side streets when the freeway stops due to severe traffic.

If a bypass diode breaks, there will permanently be a loss of voltage from the group of bypassed cells at open-circuit voltage, since the bypass is usually broken in a permanently bypassed position.

TEMPERATURES AND PV

When it gets cold, PV voltage goes up; when the solar cells get hot, PV voltage goes down. Since PV is always tested at 25C, we need to know what happens when the temperatures change, and we can do this by using PV module temperature coefficients or correction factors.

If you had to remember a number to use for temperature coefficients, I recommend estimating that the temperature coefficient **Voc is about one-third of a percent per degree Celsius** and the temperature coefficients for **Vmp and power are just less than half of a percent per degree Celsius**.

Typical temperature coefficients for crystalline PV

Temperature coefficient Voc ranges –0.3%/C to –0.35%/C
Temperature coefficient Vmp ranges –0.45%/C to –0.5%/C
Temperature coefficient power ranges –0.45%/C to –0.5%/C

The lower the coefficient, the better.

Often the module manufacturer does not give you the temperature coefficient for Vmp; when they do not give it to you, use the temperature coefficient for power.

STRING SIZING EXERCISES

We will now use the information from the module manufacturer's datasheet, the inverter manufacturer's datasheet and low-temperature data from the Solar America Board of Codes and Standards website (www.solarabcs.org) to determine how many modules can go in series.

From Solar America Board of Codes and Standards:
Low design temperature: −6C

From module manufacturer's datasheet:
Temperature coefficient Voc: −0.33%/C
Open-circuit voltage (Voc): 37V

From inverter manufacturer's datasheet:
Inverter maximum input voltage: 600V

STEP 1: CALCULATE DELTA T (DIFFERENCE IN TEMPERATURE FROM STC 25C)

$$-6C - 25C = -31C$$

Really all we need to know in step 1 is that −6C is 31C different than 25C. If the low temperature is below zero then add 6 to 25 to get 31.

If the low temperature were above zero, then we would have subtracted it from 25 and had a delta T of less than 25C. There is no place on Earth where the low temperature does not get below 25C.

I often do my string sizing calculations and do not worry about writing down the negative number for the delta T. I just have to remember that when it gets colder, my voltage will go up. Never forget what direction your calculations should go. Always use common sense every step of the way rather than getting lost in strict formulas. Think of what you are trying to accomplish.

STEP 2: MULTIPLY DELTA T BY TEMPERATURE COEFFICIENT OF VOC

$$-31C \times -0.33\%/C = 10.23\% \text{ increase in Voc}$$

Our Voc is going to increase by just more than 10% when it is −6C.

STEP 3: TURN THE PERCENTAGE INTO A DECIMAL

$$10.23\%/100\% = 0.1023$$

STEP 4: TURN THE DECIMAL INTO A TEMPERATURE CORRECTION FACTOR

$$1 + 0.1023 = 1.1023$$

If you multiply the Voc by the temperature correction factor, you will get a 10.23% increase in Voc here. Adding the 1 before you multiply ends up saving you a step in the long run. Otherwise if we multiplied the Voc of 37V by 0.1023 and got 3.79V, we would then have to add 3.79V to 37V to get our cold-temperature voltage.

NEC table 690.7 gives temperature correction factors for Voc. We are instructed by the NEC to use that table if we do not have the data from the manufacturer.

STEP 5: MULTIPLY THE TEMPERATURE CORRECTION FACTOR BY THE VOC

$$1.1023 \times 37V = 40.8Voc \text{ at } -6C$$

STEP 6: DIVIDE INVERTER MAXIMUM INPUT VOLTAGE BY COLD TEMPERATURE VOC

$$600V/40.8Voc = 14.7 \text{ in series}$$

Since we cannot divide a PV module into seven-tenths, we have to round-down to 14 in series in order to avoid going over voltage.

The answer to the question above is **14 modules in series max** under these conditions.

This was drawn out long form in order to make a point; however, let us abbreviate the above calculation. If you practice this enough you can do these calculations reliably in under 30 seconds.

SHORTCUT CALCULATIONS FOR THE EQUATION ABOVE (THE WAY I DO IT FAST)

$$6C + 25C = 31C$$
$$31C \times 0.0033 = 0.1023$$
$$0.1023 + 1 = 1.1023$$
$$1.1023 \times 37V = 40.8V$$
$$600V/40.8V = 14.7$$

14 in series max!

(In the preceding example, using the fast method, I moved the decimal of the temperature coefficient Voc two places to the left so I did not have to later change the percent to a decimal.)

Let us do a few more of these calculations until you get it down! Getting this down, so you can do it fast, will save you a lot of time on an exam.

STRING SIZING PRACTICE QUESTION 1

- Low temperature: −10C
- Temperature coefficient Voc: −0.3%/C
- Voc: 44V
- Inverter max. input voltage: 550V

Here are the calculations using the fast way:

$$10C + 25C = 35C$$
$$35C \times 0.003 = 0.105$$
$$0.105 + 1 = 1.105$$
$$1.105 \times 44V = 48.62V$$
$$550V/48.62V = 11.3$$

11 in series max!

STRING SIZING PRACTICE QUESTION 2

- Low temperature: 10C
- Temperature coefficient Voc: −0.31%/C
- Voc: 36V
- Inverter max. input voltage: 600V

Fast method:

$$25C - 10C = 15C$$
$$15C \times 0.0031 = 0.0465$$
$$0.0465 + 1 = 1.0465$$
$$1.0465 \times 36V = 37.7V$$
$$600V/37.7V = 15.9$$

15 in series max!

In the question above, the low temperature was above zero, so we subtracted it from 25 instead of adding it using the fast method.

STRING SIZING PRACTICE QUESTION 3

- Low temperature: –40C
- Temperature coefficient Voc: –0.32%/C
- Voc: 36.5V
- Inverter max. input voltage: 1000V

Fast method:

$$25C + 40C = 65C$$
$$65C \times 0.0032 = 0.208$$
$$0.208 + 1 = 1.208$$
$$1.208 \times 36.5V = 44.1V$$
$$1000V/44.1V = 22.7$$

22 in series max!

Keep going, repetition imprints it on your brain.

STRING SIZING PRACTICE QUESTION 4

- Low temperature: 19C
- Temperature coefficient Voc: –0.33%/C
- Voc: 37.5V
- Inverter max. input voltage: 750V

Fast method:

$$25C - 19C = 6C$$
$$6C \times 0.0033 = 0.0198$$
$$0.0198 + 1 = 1.0198$$
$$1.0198 \times 37.5V = 38.2V$$
$$750V/38.2V = 19.6$$

19 in series max!

STRING SIZING PRACTICE QUESTION 5

- ' Low temperature: –35C
- Temperature coefficient Voc: –0.34%/C
- Voc: 37.3V
- Inverter max. input voltage: 500V

Fast method:

$$25C + 35C = 60C$$
$$60C \times 0.0034 = 0.204$$
$$0.204 + 1 = 1.204$$
$$1.204 \times 37.3V = 44.9V$$
$$500V/44.9V = 11.1$$

11 in series max!

One of the tricks to getting really fast is leaving everything on the calculator, so that you do not have to write anything down or round any numbers. The one tricky part is when you have the cold temperature Voc in the display of your calculator and you need to divide the inverter maximum input voltage by the cold Voc. The trick here is to know how to use the 1/X button on your calculator. If I have 44.9V on my calculator display and push 1/X, then I multiply by 500V, it gives me the same thing as dividing 500V by 44.9V.

1/X calculator trick

Anytime you have a number on your calculator screen and you want to divide another number by that number, just press the 1/X button and then multiply by the number that you want your number divided into. Try it now, it works!

If you are comfortable with your cold temperature string sizing, then you can try some hot temperature string sizing; however, if you are not yet comfortable with your cold temperature string sizing, reading this will just confuse you more. If you are not ready and you need a change of pace, skip the hot temperature string sizing and come back to the cold temperature string sizing later. From my experience it takes most people a few days to grasp string sizing, so do not feel bad if you have to assimilate this new information while you sleep (learning).

HOT TEMPERATURE STRING SIZING

When it gets hot, your PV voltage will go down, and when the voltage is too low, your inverter will underperform or go to sleep. There are a few differences when we do hot temperature string sizing.

When it is hot, and the voltage is low, it is not a safety issue, so we are not concerned with the Voc. Voc is off, and when we are talking performance we want the inverter to be on at Vmp.

When we were doing cold temperature calculations we were rounding down at the end so we did not go over voltage. With hot temperature string sizing we are going to round up so our voltage does not get too low and drop out of the permissible voltage window.

What is too low? Since we are talking efficiency, we can have different opinions on this, but most people say that we do not want our source circuits to get below the low end of the MPPT range, which we can find on the inverter datasheet.

The temperature coefficient for Vmp is also greater than the temperature coefficient for Voc and we can use the temperature coefficient of power if the temperature coefficient of Vmp is not available.

We need to determine that the solar cell temperature is going to be considerably hotter than the ambient temperature. Many solar experts add 30C to the 2% hot ambient temperature from SolarABCS.org.

If we are inaccurate with our hot temperature calculations, the worst that can happen is our system turns off on a hot day. If this happens once per year for an hour, no big deal; however, as time goes on voltage will degrade slightly and this can lead to more hours per year of bad performance years later.

It is best practice to have strings as long as possible when there are no other factors, such as limited roof space or budget causing you to use a low-end "short" string.

Here is a list of what we use to calculate hot temperature short string sizing:

1. hot PV cell temperature
2. temperature coefficient Vmp
3. Vmp
4. inverter low MPPT voltage.

Since we are experts at doing the cold temperature string sizing, the math here will be easy and we can focus on the differences.

STRING SIZING PRACTICE QUESTION 6

- High cell temperature: 60C
- Temperature coefficient Vmp: −0.48%/C
- Vmp: 30V
- Inverter low MPPT voltage: 200V

> 60C − 25C = 35C (we are given the cell temperature, not ambient)
> 35C × 0.0048 = 0.168 (this time we lose 16.8% of our voltage)
> 1 − 0.168 = 0.832 (if we lose 16.8%, we keep 83.2% of our Vmp)
> 0.832 × 30V = 24.96V
> 200V/24.96V = 8.01

9 in series minimum!

In this case, since we are so close, we can make an engineering decision to round down if we have a good reason to, because it would not be a danger or code violation to have our voltage too low. Worst-case scenario with eight in series is that the system will not produce efficiently or will turn off when it is hot.

STRING SIZING PRACTICE QUESTION 7

- High cell temperature: 65C
- Temperature coefficient Vmp: −0.47%/C
- Vmp: 31V
- Inverter low MPPT voltage: 250V

Fast method:

> 65C − 25C = 40C (delta T from 25C)
> 40C × 0.0047 = 0.188 (18.8% lost)

$$1 - 0.188 = 0.812 \ (81.2\% \ kept)$$
$$0.812 \times 31V = 25.17V$$
$$250V/25.17V = 9.9$$

10 in series minimum!

STRING SIZING PRACTICE QUESTION 8

- High *ambient* temperature: 35C
- Temperature coefficient Vmp: –0.46%/C
- Vmp: 32V
- Inverter low MPPT voltage: 400V

Fast method:

$$(35C + 30C) - 25C = 40C \ (added \ 30C \ to \ compensate \ for \ cell \ above \ ambient)$$
$$40C \times 0.0046 = 0.184 \ (18.4\% \ lost)$$
$$1 - 0.184 = 0.816 \ (81.6\% \ kept)$$
$$0.816 \times 32V = 26.11V$$
$$400V/26.11V = 15.3$$

16 in series minimum!

STRING SIZING PRACTICE QUESTION 9

- High ambient temperature: 33C
- Temperature coefficient Vmp: –0.44%/C
- Vmp: 31V
- Inverter low MPPT voltage: 350V

Fast method:

$$(33C + 30C) - 25C = 38C$$
$$38C \times 0.0044 = 0.1672 \ (16.72\% \ lost)$$
$$1 - 0.1672 = 0.8328 \ (83.28\% \ kept)$$
$$0.8328 \times 31V = 25.8V$$
$$350V/25.8V = 13.6$$

14 in series minimum!

Calculator shortcut

When we get the calculation in which we have the percent voltage lost in the decimal form and need to subtract that from 1, we can just subtract 1 from the decimal that represents the lost percentage. We end up with negative numbers on the calculator that we can ignore. If you can ignore a negative number, you can save time. If we want to see a positive number we can press the +/- button.

From string sizing question 8:

$$0.184 - 1 = -0.816$$
$$-0.816 \times 32 = -26.11$$
$$400V/-26.11 = -15.3$$

So, 16 in series

I just ignored the negative, which can be interpreted as optimistic.

STRING SIZING PRACTICE QUESTION 10

- High cell temperature: 67C
- Temperature coefficient Vmp: −0.45%/C
- Vmp: 30.5V
- Inverter Low MPPT Voltage: 300V

Fast method:

$$67C - 25C = 42C$$
$$42C \times 0.0045 = 0.189 \ (18.9\% \ lost)$$
$$1 - 0.189 = 0.811 \ (81.1\% \ kept)$$
$$0.811 \times 30.5V = 24.7V$$
$$300V/24.7V = 12.1$$

13 in series minimum!

You are now officially a string-sizing expert! Remember, though, that the proper NEC name for a string is a PV source circuit. The Europeans call it a string in their codes and standards.

In order to become an expert in all that is parallel in PV, you should know that you do not connect PV source circuits (strings) in parallel that are of different lengths or different types of modules. All strings paralleled together at a combiner (where the parallel connection happens) should be of the same voltage.

Parallel PV connections are those in which all of the negatives are tied together and all of the positives are connected together, which increases current and not voltage. Parallel connections happen at a combiner box and the parallel connections increase current.

Series is connecting positive to negative to increase voltage, which does not increase current.

A series string with 1000 solar cells in series will have the same current as a single solar cell.

It is acceptable if we parallel different source circuits together that have different orientations. We never should have different orientations within a single source circuit.

If we have different orientations within a source circuit, then whatever module has the least amount of light will limit the flow of current throughout that string.

If we have strings of different lengths connected together at a combiner, there will be different IV curves with different maximum power points connected together which can only operate at a single voltage, so it will be impossible for different string lengths to both MPP together.

Many modern inverters have multiple MPP inputs and in those cases we can have different lengths of source circuits on different MPP inputs. If a combiner has different MPP inputs, then strings can be of different lengths. Every time we go through smart electronics, it is a game changer (and one more thing to break).

If module-level electronics are used, source circuit rules do not follow. It is incorrect to call modules connected together with power electronics (power optimizers) PV source circuits or strings, according to most experts. Microinverter circuits are definitely not source circuits. We could, however, say that microinverters have a string of 1 and some people will do the calculations and string size a module for an inverter. I did a string sizing calculation for a microinverter located on the south pole after seeing a photo of microinverters tilted at latitude on the south pole. As I suspected, the modules would have been over voltage during a cold snap.

Once PV is connected to electronics, the electronics control the voltage and current, so the calculations are not the same as for conductors that are directly connected to silicon solar cells.

Silicon PV is really a magic crystal that can turn light into electricity with no moving parts. Ninety percent of the Earth's crust is made from materials that contain silicon; PV is a power plant for everyone and that is revolutionary!

PV system components

Chapter 5

Besides knowing about PV, we have to understand the components required, how to configure the PV system and how these system components interact with each other. System components that are important for a solar salesperson are many different types of inverters, power optimizers, charge controllers, batteries, conductors, conduits, cables, fuses, circuit breakers, combiners, disconnects, busbars, load centers, meters, etc.

First, we will look at a simple single line diagram (SLD) of a utility interactive system, which is by far the most common type of PV system in the world today.

Figure 5.1 shows an SLD of a very simple utility interactive PV system. There are other components that we could have put in the drawing, but for simplicity's sake we have just included the major components.

Some other components that we could have included are disconnects, combiner boxes, conductors and conduit. Disconnects are added for safety so we can

Figure 5.1 Utility interactive PV system with load-side connection.

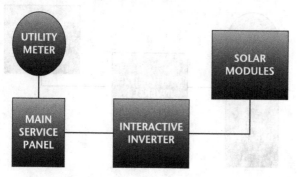

isolate different system components. If the disconnect is a load break rated disconnect, then it can also be used to turn the system off. Non-load break rated disconnects can catch fire if they are used to turn off the system when it is on and there is current flowing through the disconnect. Sometimes we refer to disconnects as isolating devices. Even the connectors between the modules are isolating devices and/or non-load break rated disconnects.

In the single line diagram above, the PV is connected to the main service panel, which is called a load-side connection. If the PV system were connected between the main service panel and the meter it would be called a supply-side connection.

When we talk about supply-side (line-side) connections and load-side connections, we are referring to the load or the line side of the main breaker. Every breaker has line and load sides. Line or supply is always the utility and load is always the solar side. This is sometimes confused, since we are sending power from the load side, which is unusual. When we feed more power on the load side of the main breaker, we are adding more current on the load side than the main breaker was originally protecting the load side from. It can be tricky calculating the load-side connection inverter sizes, especially after the 2014 NEC was released.

A supply-side connection is usually more expensive, but we can use a larger inverter than with a load-side connection, since we are not worried about adding more currents to the load side of the main breaker. With a supply (line) side

Figure 5.2 Utility interactive PV system with supply-side connection.

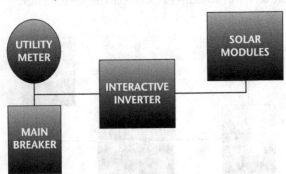

connection the main breaker will protect everything on the load side, just as it always has before the PV came along.

The typical interactive inverter will get its input from the PV modules and all of the conductors on the dc side of this inverter will be sized based on doing different calculations and corrections based on the short circuit current rating of the PV module. Electricians and engineers are used to doing wire sizing based on operating currents and they are also used to short circuit currents that are many times greater than operating currents. PV, on the other hand, has a short circuit current that is about 7% greater than current at maximum power (Imp) under standard test conditions (STC). Also, with PV, the short circuit current rating is done at STC, and with PV we will operate above the STC ratings at times due to increased irradiance (brightness), which increases current, and decreased temperatures, which increase voltages.

When we learned about string sizing in the last chapter, it was based on the maximum input voltage and the input low MPP voltage of the inverter. The circuits on the output side of the interactive inverter are based on the output current rating of the inverter and the voltage of the utility for utility interactive inverters. When we size wires on the ac output of an inverter, we do not consider anything on the dc side of the inverter.

A combiner box will combine PV source circuits into a PV output circuit, and when there are more than two PV source circuits to be combined, we will need fuses in the combiner box. On inverters larger than 25kW, we often have combiner boxes separate from the inverter; on systems smaller than 25kW we often have the combiner with the inverter. When it is with the inverter, it can look like it is part of the inverter, but usually it is separable in North America. Combiner boxes usually come with dc disconnects. If you see a good price on a combiner box without a dc disconnect, beware, because it will require a separate dc disconnect according to the NEC, which will drive up your costs.

Combiner boxes have fuses that have to be rated for the dc current and voltage which will be used. The voltage we will get from the cold temperature maximum system voltage calculations that we did with our string sizing calculations in the previous chapter; the current of the fuse is usually 15A and can be calculated by multiplying short circuit current of the module by 1.56 and rounding up to the next common fuse size, which usually comes out at 15A. Make sure that your combiner fuses are rated for the correct dc voltage!

Figure 5.3 Grid-tied PV system three-line diagram.

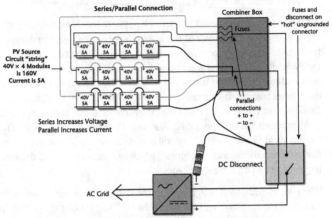

In a negative grounded system, inside inverter there is a connection between negative and ground.
If more than one amp goes through this connection, there is a ground fault and device deactivates inverter.

In Figure 5.3 we can see positive and negative circuits, which means that this is not a single line diagram, so we call it a three-line diagram.

This system is negatively grounded through a ground fault detection and interruption fuse inside the inverter, which is represented by the fuse by the negative pole of the inverter. A benefit of having a grounded inverter is that we do not put fuses on the negative (grounded conductor) and we do not disconnect the negative (grounded conductor) at the disconnect.

Many of the newer inverters are ungrounded (transformerless) and **do require fuses** on the positive and the negative and you **do have to disconnect the positive and the negative** when turning the system off. There is talk that the Code may change someday and that ungrounded inverters may have to only disconnect one pole at the disconnect and only have to fuse positive or negative.

We can see the combiner box and the dc disconnect are separate from the inverter in this diagram. We often make the diagram simple by separating the components, even when they are in the same box. In reality, this inverter with three PV source circuits would probably come with a combination dc disconnect/combiner that attaches right to the inverter. Outside of North America, in most other countries, the PV connectors will come straight to the inverter, which is simple, quick, cheap and not NEC compliant.

CONDUIT AND CONDUCTORS

Conduit is a type of a raceway and is a pipe that will protect wires. Some conduit is plastic (PVC) and other conduit can be steel or aluminum. A very common type of conduit used for PV systems in the USA and Canada is steel electrical metallic tubing (EMT), which is relatively cheap, but takes skill to bend around corners of a building and install. Experienced electricians are EMT artists.

When PV source and output circuits are run through a building, they have to be inside a metal raceway or metal-clad cable up until the first readily accessible dc disconnect. We are allowed to have plastic conduit on the outside of the building, which is common in Hawaii, where corrosion is widespread. Plastic typically will not last as long as steel EMT. In most of the USA, EMT is used for circuits inside and outside of the building. Plastic will not last as long in the sun in most places. Hawaii is an exception.

Conductors used are often 90C rated because of the heat that PV system cables are exposed to. Many conductors used for wiring in the USA are 75C rated, but when we are working with PV, which is in the sunlight, it is strongly recommended to use 90C rated wire.

The NEC has various tables used to size wires. Some of the tables will tell us how much current a wire can handle under normal indoor 30C conditions. The tables can differentiate depending on whether the wires are single conductors in air or if they are grouped together or inside a raceway (conduit).

When it is hotter than 30C, there is a table that can be used to derate the conductors in the tables mentioned above, so we can see how much current they can handle under hotter conditions.

Also, if conductors are above a roof in a raceway in the sunlight, we have a table that will give us a temperature to add to the ambient temperature. If sunlight hits a pipe on a rooftop, it can heat it up like a solar thermal cooker.

Another table is one that we use to derate for more than three current-carrying conductors in conduit. If there are a lot of conductors in conduit, they will heat up even more, and be unable to dissipate the heat so they will not be able to carry as much current.

When wires heat up due to currents, sunlight, too many wires in a conduit, etc., we have to increase the wire size so the wire does not heat up too much from the current going through the wire.

Complete wire sizing is very difficult and not in the scope of this book, the PV Technical Sales Exam or the job of a solar PV salesperson. Wire sizing is in the scope of the NABCEP PV Installation Professional Exam and the book, *PV Engineering and Installation* by Sean White.

Voltage drop is another consideration that we will have to bear in mind, and if we are selling a system where there are long distances or lower voltages, we will have to spend more money on thicker conductors. The way we size wires in the USA and many other parts of the world is the American Wire Gauge (AWG) system, where the larger the number, the smaller the wire. Then when we get to size zero, our numbers will increase by adding zeros. Here is an example of increasing wire sizes from smallest to largest:

14 AWG, 10 AWG, 6 AWG, 1 AWG, 1/0 AWG, 4/0 AWG

From the sizes above, we can see that 4/0 is larger than 1/0. 4/0 means four zeros of 0000 and to get larger than 4/0 we have to state the cross-sectional area in kcmil.

BATTERIES AND ENERGY STORAGE

Batteries with PV is a thing of the past and a thing of the future. In the past, almost all PV systems were off-grid battery-based systems. In the future, as more PV is connected to the grid, we will need to have batteries to store the energy and to condition the power.

There is the old-style off-grid system, which was dc coupled and had PV charging batteries through a charge controller, and there is a new-style (more expensive) ac coupled system where an interactive (grid-tied) inverter works with a battery inverter/charger that can also work backwards and charge the batteries with the power that the interactive inverter is making. These ac coupled systems work by using a battery inverter to make a "micro-grid" that will fool the interactive inverter into thinking that the grid is there and it will work. If the inverters in the ac coupled system are from the same company and can communicate with each other, then they can have extra abilities. If they are not from the same company or do not have the communication abilities, the battery inverter/charger can turn off the interactive inverter by shifting the frequency of the micro-grid slightly to make the interactive inverter anti-island (turn off). A benefit of ac coupling is that it can work using standard interactive inverters. Interactive inverters are most popular and the technology is more mass-produced, more researched and more efficient. Also, the interactive inverters usually have higher voltage PV source circuits, which is good for saving wire costs and going over longer distances.

On the other hand, old-style dc coupled systems typically have lower voltages (requiring larger wires and having more losses) and they are typically less efficient than the more expensive and complicated to operate and install ac coupled systems. DC coupled systems are much more cost-effective for smaller PV systems.

Utilities are starting to have policies through which they will pay for people to assist the grid with batteries, and it is a new and exciting time, with the prices of batteries expected to drop drastically as production increases to fulfill needs. Without a government or utility policy that will pay people to assist the grid with batteries, batteries will drastically increase the cost of a PV system without a lot of benefit. When there is no policy, the customer often says that they want batteries, until they find out that it may cost twice as much for a system with battery backup, depending on how much backup they want.

There is another type of inverter that has what is called a secure power supply; this inverter will power a single circuit when the utility is down without batteries. When the sun sets or when your power requirements are more than the system can produce, the secure power supply will shut off. This type of system has power when the grid is down and is not nearly as expensive or complicated as a battery system.

There are many new systems that have not been released that people are talking about. I would caution you to beware of big talk with nothing to show. In this fast-growing industry, there are many companies making a lot of money selling hype. Have you ever heard of solar roadways or BIPV? It sounds great, but it is not going to get a return on investment that customers and banks require.

If someone has the extra money, that is great! Sell them a BIPV lithium battery backup system with ten days of storage for their air conditioners. Their PV system might cost as much as a house today, but if they want to spend a lot on renewable energy, be their conduit!

Utility interconnections Chapter 6

It is important for someone selling a grid-tied PV system to know where and how the system will be connected to the grid. There are many different ways to connect a PV system to the grid; some of them easy and other ways more complicated. There have also been a lot of changes regarding interconnections between the 2011 and the 2014 versions of the National Electric Code.

In this book, we will focus on the most common ways PV systems are connected and try not to get lost in the details of feeder tap connections, as solar salespeople will undoubtedly work with engineers when contemplating the most difficult of connection calculations. We will, however, make it clear that there are different opportunities for making connections, which we did not have in the 2011 code.

Two basic types of interconnections we have are load side and supply side. When we are talking about load in line in this context, we are referring relative to the main service disconnect. This is typically a main breaker in a service panel or a main breaker feeding a service panel. Line side of the main breaker is the utility side and load side is the other side.

SUPPLY-SIDE CONNECTION

The supply (line) side of the main breaker is called a supply-side connection. Historically, many solar installers used to call this a line side tap, which is an incorrect term but is still used in the field today. A supply-side connection is not a tap and the tap rules will apply to other connections that we can make on the load side of the main breaker.

With a supply-side connection, you can typically add as much PV as you would ever need to in a net-metering situation. We are allowed to have the breakers on a supply-side connection(s) add up to the size of the service. That means that if we had a 200A service, we could have a 200A inverter overcurrent protection device, which is overkill and much more than the person would need in a net metering situation.

The reason why we are safe adding a lot of PV on the supply-side more so than the load side is that on the supply side we are on the utility side and supplementing the utility, which we are already protected from by the main breaker. If we were on the load side, we would be adding extra current on the load side in addition to the current that is coming through the main breaker.

LOAD-SIDE CONNECTIONS

With load-side connections, we have to be careful, since we are adding more current than the main breaker on the load side of the main breaker. Since the main breaker is protecting the load side of the busbar and equipment, then we can have trouble if we supplement this current with too much solar PV produced current in the wrong place.

THE 120% RULE

The most popular way to interconnect solar systems is with a load-side connection is using the 120% rule. The 120% rule states:

> When an inverter and the utility are connected on **opposite sides** of a busbar, 125% of the inverter current plus the rating of the main breaker cannot exceed 120% of the ampacity of the busbar.

The reason why we can exceed the busbar ampacity at all is because if the solar and the main breakers are on opposite sides of the busbar, then there will not be a case where there will be overcurrents on any spot on the busbar. If the solar and the main breakers were adjacent to each other, it would be like having a big breaker that is the equivalent of the size of both breakers together. If we separate the breakers they will go to loads before the currents will get a chance to add up.

In the 2011 NEC and earlier, we used the rating of the inverter overcurrent protection device in our calculations for the 120% rule. After the 2014 NEC, we use 125% of the inverter current instead. Inverter overcurrent protection devices are based on 125% of the inverter current and then rounding up to the next common overcurrent protection device. The newer way gives us more in some cases, because we are not penalized for the rounding up effect.

Figure 6.1 120% rule: main breaker and solar breaker on opposite sides of the busbar.

Here is the equation for the 120% rule:

$$\text{busbar} \times 1.2 \geq \text{main breaker} + 125\% \text{ inverter current}$$

And we can derive the inverter size from the equation:

$$125\% \text{ inverter current} \leq (\text{busbar} \times 1.2) - \text{main breaker}$$
$$\text{inverter current} \leq ((\text{busbar} \times 1.2) - \text{main breaker}) / 1.25$$

If we want to solve for inverter power, we will multiply by the ac voltage. In this example we will use the standard US residential inverter voltage of 240V:

$$\text{inverter power} \leq (((\text{busbar} \times 1.2) - \text{main breaker})/1.25) \times 240V$$

And there we have it in one neat equation.

Table 6.1 shows common busbar and main breaker combinations with corresponding maximum inverter sizes at 240V ac.

Table 6.1 120% rule common breaker and inverter combinations

BUSBAR	MAIN BREAKER	125% OF INVERTER CURRENT	INVERTER CURRENT	MAX. INVERTER POWER AT 240V
100A	100A	20A	16A	3.84KW
125A	100A	50A	40A	9.6KW
200A	200A	40A	32A	7.68KW
225A	200A	70A	56A	13.44KW

Understanding the 120% rule is crucial for any good solar salesperson.

SAMPLE 120% RULE QUESTION

If you walk up to a house and see that the main breaker is 200A and the busbar is 225A, what would be the largest inverter you could connect?

First we are going to multiply the busbar by 1.2 (120%):

$$225A \text{ busbar} \times 1.2 = 270A \text{ allowance}$$

Then we subtract the 200A main breaker from the 270A allowance:

$$270A \text{ allowance} - 200A \text{ main breaker} = 70A$$

70A is 125% of inverter current, so we can divide 70A by 1.25 to get our inverter current:

$$70A / 1.25 = 56A \text{ inverter}$$

(multiplying by 0.8 is the same as dividing by 1.25).

Since US residential services are 240V and volts × amps = watts:

$$240V \times 56A = 13,440W$$

Therefore, our answer is that the largest load-side inverter that we can add is going to be 13,440W or 13.4kW.

We have to be sure that we put the inverter breaker on the opposite side of the busbar from the main breaker. If the main breaker is not on a far end of the busbar, then we cannot apply the 120% rule.

100% RULE

If the sum of the main breaker plus 125% of the inverter current does not exceed the busbar ampacity, then we do not have to put the inverter and main breakers on opposite sides of the busbar – we can put the breakers anywhere on the busbar.

It gets confusing when we talk about the 120% rule or the 100% rule and then we talk about 125% of the inverter current. There is no 125% rule that we talk about here. 125% of the inverter current is also described as required ampacity for continuous current. Throughout the Code, when there is something that has the ability or potential to operate for three hours or more, we have to multiply the current by 125% for continuous current as a safety factor. All of the equipment in a PV system needs to be able to handle at least 125% of maximum circuit currents.

Maximum circuit current for the ac output of an inverter is written on the label or calculated by power / voltage = current.

Continuous current for the ac output of the inverter is the current above multiplied by 1.25.

Maximum circuit current for a PV module is Isc × 1.25 and continuous current for a PV module is Isc × 1.56.

FEEDER CONNECTIONS

There are rules for connecting to feeders that were added to the NEC in 2014. An example of a feeder is a conductor that goes from a main service panel to a subpanel.

There are many ways to connect to feeders and some of the connections can get very complicated. Do not get overwhelmed; realize that, in practice, you may be working with a company engineer to sell this system. If the rest of this section is overwhelming, you can skim over it and be aware that there are other options.

As we can see in Figure 6.2, if we connect an inverter to a feeder, we will be exposing the load side of the feeder to the currents from the feeder breaker and from the inverter. We have to make sure that the feeder on the load side of the inverter connection point does not get exposed to unsafe currents. There are two main ways of accomplishing this.

Figure 6.2 Feeder connection problem.

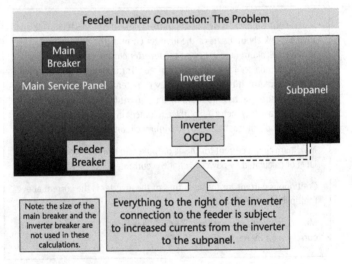

PROTECTING FEEDER ON LOAD SIDE OF INVERTER CONNECTION POINT WITH CONDUCTOR SIZE

Solution A is making sure the feeder on the load side of the inverter connection point has an ampacity of at least the feeder breaker plus 125% of the inverter current, as shown in Figure 6.3.

Figure 6.3 Feeder conductor on load side of inverter connection point sized to handle 125% of inverter current plus feeder breaker rated current.

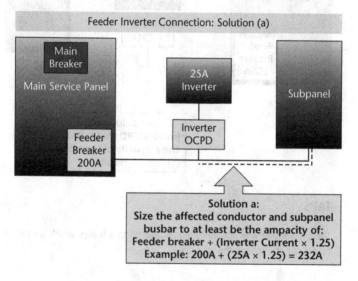

PROTECTING FEEDER ON LOAD SIDE OF INVERTER CONNECTION POINT WITH OVERCURRENT PROTECTION DEVICE

Solution B is having an overcurrent protection device between the inverter connection point and the subpanel, as shown in Figure 6.4.

Figure 6.4 Feeder on load side of inverter connection point protected with overcurrent protection device.

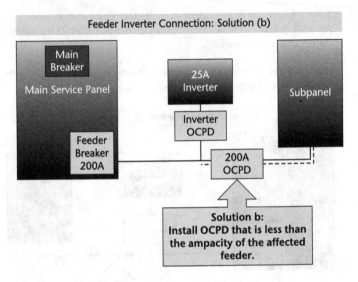

TAPS

There is another way of connecting an inverter to a feeder, which is called a feeder tap. This follows the tap rules.

10-FOOT TAP RULE

The 10-foot tap rule states that if you are connecting to a feeder and your conductor going to the feeder is less than ten feet, then your conductor going from the inverter to the feeder has to be at least 10% of the sum of the feeder breaker plus 125% of the inverter current. Figure 6.5 makes this simpler.

Figure 6.5 10-foot tap rule.

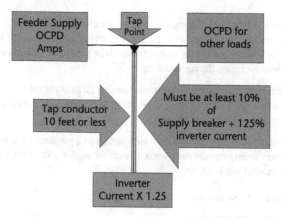

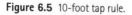

25-FOOT TAP RULE

The 25-foot tap rule states that if you are connecting to a feeder and your conductor going to the feeder is between 10 and 25 feet, then your conductor going from the inverter to the feeder has to be at least one-third of the sum of the feeder breaker plus 125% of the inverter current. Figure 6.6 makes this simpler.

Figure 6.6 25-foot tap rule.

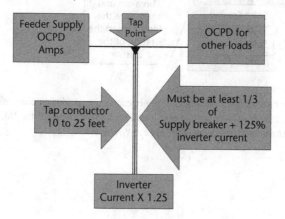

SUM RULE

There is another way of doing a load-side connection and I like to call it the sum rule, because you sum up all of the breakers on the busbar and if the amps do not exceed the rating of the busbar, then you can put as much solar as you want without exceeding the busbar rating.

This type of interconnection will require adding up all of the solar and the load breakers. Also, a label will have to be placed so nobody in the future will come along and add too many breakers.

The sum rule is often used to combine inverters on a subpanel, since most load centers already contain more breakers than the busbar is rated for. Figure 6.7 shows what this looks like.

Figure 6.7 The sum rule.

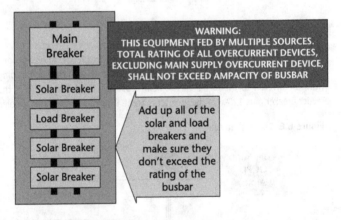

For more detailed explanations of interconnections, including code references, read *Solar PV Engineering and Installation, Preparation for the NABCEP PV Installation Professional Exam*, by Sean White.

70-question practice exam

Chapter 7

This is a 70-question practice exam designed to prepare you for the real NABCEP Solar PV Technical Sales Exam or the real world.

If you would like to gauge how you will do on the exam right now and get the best practical experience possible, sit down for four hours with a Casio fx260 calculator and a pencil and see how you do on this exam.

Following this exam is Chapter 8, which contains detailed explanations for all of the questions, and is the heart of this book.

Set your clock, ready, set, go. . .

1. A south-facing roof has measurements of 91 feet north to south and 154 feet east to west. What is a rough estimate of how much PV you can fit on the roof?

   a. 140kW
   b. 15kW
   c. 500kW
   d. 90kW

2. Which are the best tools for measuring solar access?

   a. Tape measure, pyranometer and transit
   b. Pyrheliometer, temperature sensor, calculator
   c. Sunpath calculator, compass
   d. IV curve tester, compass

3. Which of the following is a PBI (performance based incentive)?

   a. Investment tax credit (ITC)
   b. Depreciation
   c. Feed-in-tariff (FIT)
   d. Time of use (TOU)

4. Southeast- and southwest-facing roofs are going to have solar on them. Each rooftop will fit seven 250W modules on it. What would be the best inverter combination for this project? Assume that the given PV source circuit sizes will adequately configure with the inverter.

   a. 1 PV source circuit of 14 modules in series on a 3.8kW inverter with 2 MPPs

   b. 2 PV source circuits of 7 modules each on a 3.5kW inverter with a single MPP

   c. 2 separate 2kW inverters

   d. A single 3kW inverter with 2 MPPs

5. On a residential two-story composition asphalt shingle roof, you discovered that you can fit more modules by orienting them in landscape mode and you are using a standard rail system. The rail instructions say that the rail must be mounted perpendicular to the rafters, which go from the ridge to the eaves. What must be taken into consideration before mounting the modules?

   a. Do the manufacturer's instructions let you mount the modules by connecting to the short edges?

   b. Are the rafters 24" on center?

   c. Do the rails have splice kits that allow for expansion?

   d. Use three rails per module.

6. In Calgary, Alberta, there is a tendency for snow, wind and rain. A big-box retail store would like to have as much solar PV as possible installed on the roof. In order to not waste time on a job that may never happen, what would be the best course of action?

   a. Avoid snowdrift by mounting the system with an increased tilt angle.

   b. Hire a structural engineer to determine whether the roof can take the weight of the PV system.

   c. Design a ballasted system with wind deflectors to avoid extra ballast weight.

   d. Design a penetrating system.

7. Who will approve seismic reinforcements on a commercial rooftop ballasted PV system job in a known earthquake fault zone?

   a. Racking manufacturer
   b. Module manufacturer
   c. Inverter manufacturer
   d. Building department

8. You are selling and designing a 100kW array that has a single inverter that has seven modules shaded about two hours per day. What is the best way to get maximum production out of this system? The PV source circuits are ten modules in series.

   a. Have each of the seven different shaded modules on a different PV source circuit
   b. Have all of the modules on the same PV source circuit
   c. Have three shaded modules on one PV source circuit and four shaded modules on another PV source circuit
   d. Have one shaded module on one PV source circuit, two shaded modules on another PV source circuit and four shaded modules on another PV source circuit

9. If you are canvassing locations and you do not go to the house where you would like to install PV, what would be the best way to get information about the feasibility of the job?

   a. NREL Redbook data
   b. Satellite data
   c. Weather data
   d. Homeowner opinion

10. A customer has a time of use (TOU) utility rate schedule. Their off-peak rates are $0.10 per kWh, their partial peak rates are $0.12 per kWh and their peak rates are $0.15 per kWh. The customer's total energy bill for the month of August was $2000 and they used 3000kWh off-peak and spent $200 on partial peak energy. How much peak energy did they use?

   a. 1.2MWh
   b. 10,000kW
   c. 10MWh
   d. 5,000kWh

11. What is the social benefit of having a MW of PV on your rooftop?

    a. CO_2 reduction
    b. Saving oil
    c. Creating green jobs
    d. SRECS

12. When is the appropriate time to tell a cash customer about the permit costs associated with installing a PV system within their particular building department zone?

    a. When the permit package is filed
    b. At the proposal stage
    c. After the contract is signed
    d. At the filing of the interconnection agreement

13. If you were using frameless PV modules (laminates) rather than framed modules at a low tilt angle, what would be a benefit in the context of less derating for production?

    a. Lower price of modules
    b. Manufacturer's production tolerance
    c. Module mismatch
    d. Soiling

14. In a location that would get production of 1.3MWh per year for every kW of PV installed, how large of a system would we need if we were to offset all of our energy and used 14,000kWh per year?

    a. 7kW
    b. 8.1kW
    c. 9.5kW
    d. 10.8kW

15. A customer has a bill for $5000 for a month of energy with a tiered rate schedule. There are equal charges for three different tiers for two-thirds of the bill and there is a demand charge for one-third of the bill. About how much can they save by getting a PV system?

    a. $5000
    b. $3400
    c. $4000
    d. $1200

16. What is the most important factor in a good long-term return on investment for a FIT project?

    a. kWh per year
    b. kWh/square meter
    c. Tax equity
    d. Information technology

17. At what height must a worker use fall protection?

    a. 4 feet
    b. 8 feet
    c. 12 feet
    d. 6 feet

18. Who can open an electrical box according to OSHA?

    a. NABCEP Certified person
    b. Qualified person
    c. NABCEP Entry Level Certificate holder
    d. UL Certified Installer

19. What is the official term that we use to call all different forms of materials that we use to keep workers safe?

    a. Safety precaution materials
    b. Safety protective equipment
    c. Personnel protective equipment
    d. Personal protective equipment

20. What determines the maximum size of a supply-side inverter connection?

    a. 120% busbar – 125% of the inverter current
    b. 120% busbar – solar breaker
    c. The rating of the service
    d. Loads

21. What size overcurrent protection device would you use for the ac output of a 7kW inverter on a single-family house?

    a. 25A
    b. 40A
    c. 30A
    d. 45A

22. What is the voltage of a typical small commercial service?

    a.   120/240V wye

    b.   240/208V

    c.   120/208V

    d.   230/400V

23. You notice that the inverter says it is making more energy in the winter than it ever did in the summer. What is most likely happening?

    a.   Cold and wind increases production

    b.   Inverter display broken

    c.   Array was broken and wind blew it back in place

    d.   Azimuth must have been more favorable for winter production

24. An inverter can have a top voltage of 500V and the PV being used has an open circuit voltage of 36.8V. With a low temperature of –30C and a typical voltage temperature coefficient, what would be the maximum number of modules in series?

    a.   14

    b.   8

    c.   9

    d.   11

25. You are installing a large commercial rooftop PV system in stages. The first stage is 500kW, and at the end of year 1 the system makes 704MWh of energy, preventing 528 tons of carbon dioxide from going into the atmosphere. The next stage of the project is to install 258kW with the same expected performance. At the end of the year after stages 1 and 2 are installed, how much carbon dioxide will the entire system prevent from going into the atmosphere?

    a.   754 tons

    b.   595 tons

    c.   701 tons

    d.   800 tons

26. An optimally placed PV system in a location in the Nevada Desert makes 1800kWh/kWp/yr. If there were another system being installed in the same way a mile away that was 30kW, then how many MWh would you expect the system to make in 20 years?

    a. 76
    b. 54
    c. 1080
    d. 1180

27. A typical 3kW utility interactive PV system produces 8% less per year the second year than it did the first year. What is the most likely problem of the following? You checked the system and it appears to be working properly.

    a. Blown PV source circuit fuse in combiner
    b. Weather fluctuations
    c. Module performance degradation
    d. Magnetic declination

28. What is the purpose of calculating magnetic declination?

    a. To determine the longitude
    b. To determine the latitude
    c. To determine the tilt angle
    d. To determine azimuth

29. What would be the best, most cost-effective way to interconnect a 4kW STC PV system on a 120/240V residential service? The house has a typical 100A service with a 100A busbar and a 100A main breaker on the top of the busbar.

    a. With a 4kW inverter and a supply-side connection with a 25A inverter overcurrent protection device.
    b. With a 3.8kW inverter on a 20A circuit breaker with a load-side connection.
    c. With a 3.8kW inverter on a supply side connection with a 20A over-current protection device.
    d. With a 4kW inverter on a 20A circuit breaker with a load-side connection.

30. Which of the following is the largest inverter you can connect on a load-side connection with a 225A busbar and a 200A main breaker on a typical residential main service panel?

   a.   7kW
   b.   8kW
   c.   12kW
   d.   15kW

31. In a location with an insolation of 5.4 peak sun hours and a system with a derating factor of 0.77, how many kWh per year would a 10kW system produce?

   a.   15,180kWh per year
   b.   12,000kWh/kWp/yr
   c.   1400kWh per year
   d.   3750kWh

32. A residential customer is buying a net-metered PV system and they are comparing the system to an investment in the stock market. What benefit does an investment in a PV system have that an investment in green energy stocks does not have?

   a.   You are taxed on income from a PV system, but not at the capital gains rate
   b.   The taxes from owning a PV system are lower than the taxes from owning green energy stocks
   c.   A PV system is greener than green energy stocks
   d.   You are not taxed on the money saved by a PV system, but you are taxed on income from the stock market

33. You are selling a PV system on a house with a preexisting generator and the customer asks you where you will connect the PV system. He asks if the typical grid-tied PV system will work and feed back to the generator instead of the grid when the power is out. Where will you connect the PV system and will it work with the generator?

   a.   Connect the PV system inverter on the load side of the generator automatic transfer switch (ATS) and the inverter will work with the PV system when the grid is down.
   b.   Connect the inverter on the supply side of the generator ATS and the inverter will work with the generator when the grid is down.

    c.   Connect the inverter on the supply side of the generator ATS and the inverter will not work with the generator when the grid is down.

    d.   Connect the inverter on the load side of the generator ATS and the inverter will not work when the grid is down.

34. What is the best application for MACRS?

    a.   Putting PV on your own house

    b.   Putting PV on your own business

    c.   Putting PV on your non-profit

    d.   Putting PV and wind on your own house

35. If an 11kW PV system will make 14,895kWh per year, then how much energy will a similar PV system make in the same location if it consists of 48 PV modules with the following specifications? Voc = 37.5V, Vmp = 30.4V, Imp = 8.56A and Isc = 9.12A.

    a.   2010kWh per year

    b.   32,800kWh per year

    c.   1,693kWh per year

    d.   17MWh per year

36. There is a house with a 100A main utility disconnect outside of the house and the 125A service panel on the inside of the house. If planning to connect a 10kW inverter to the house, what is the best place to connect the inverter? There are 90A loads connected to the service panel, the conductor going from the main utility disconnect to the service panel is rated for 130A and you are avoiding a connection to a feeder.

    a.   Between the main utility disconnect and the service panel

    b.   On the far end of the busbar in the service panel

    c.   Between the main utility disconnect and the meter

    d.   On the supply side of the meter

37. According to the 120% rule, which is the largest inverter of the following that you can fit onto a 200A busbar with a 200A main breaker on a 240V ac inverter?

    a.   5kW

    b.   6000W

    c.   7000W

    d.   8kW

38. If a PV system were losing 1% per year due to module degradation, after 30 years what would be the annual loss due to degradation?

   a. 30%
   b. 70%
   c. 74%
   d. 26%

39. If the price of electricity goes up 3.5% per year for 20 years, what would be the price of electricity 20 years later if the price at the beginning were $0.16 per kWh?

   a. $0.34 per kWh
   b. $0.27 per kWh
   c. $0.32 per kWh
   d. $0.29 per kWh

40. If your PV module specifications were Voc = 37.2V, Vmp = 30.1V, Imp = 8.3A and Isc = 8.87A, and you could expect the PV module cell temperature to reach 65C, what would be the minimum number of modules that you could put in series if the inverter MPPT range was 200V to 480V and the inverter maximum input voltage was 600V?

   a. 7
   b. 8
   c. 9
   d. 10

41. A building has a time of use (TOU) utility rate schedule. The rates are as follows: 9 a.m. to 1 p.m. is $0.12 per kWh, noon to 7 p.m. is $0.22 per kWh and 7 p.m. to 9 a.m. is $0.10 per kWh. If a person uses an even amount of electricity throughout the day, what is the best way to face the PV modules at latitude 33N in a place without dramatic weather patterns?

   a. Southeast
   b. South
   c. Southwest
   d. West

42. A residence has a tiered utility rate schedule. The low tier is $0.15 per kWh and includes the first 250kWh per month; the next tier is $0.19 per kWh and includes the next 350kWh; and the final third tier is $0.36 per kWh and includes all remaining kWh. In January, you have used 1103kWh. What would your energy bill be?

    a. $285
    b. $103
    c. $397
    d. $94

43. If an object is five meters high and the solar elevation angle is 28 degrees, how long will the shadow be?

    a. 5.2 meters
    b. 9.4 meters
    c. 6.3 meters
    d. 3.8 meters

44. How much PV can you fit on a 44 feet wide × 32 feet north to south south-facing rooftop in portrait orientation if you are going to have a 0.5-inch space between all of the modules and if the fire department requires a three-foot perimeter on the ridge and the sides of the roof? You also need to leave a one-foot space at the bottom of the roof so that the rain will go into the gutters. The dimensions of the modules are 66 × 39 inches and they are 260 watts each.

    a. 14.3kW
    b. 44kW
    c. 10.2kW
    d. 47kW

45. Which is the best type of inverter for designing a system where you can add more modules later?

    a. String inverter
    b. Central inverter
    c. Microinverter
    d. Transformerless inverter

46. What would usually be a better investment?

    a. Installing thin-film PV
    b. Installing crystalline PV
    c. Switching from incandescent to CFL lighting
    d. Savings account

47. Can you mount an inverter with the top of the inverter four inches from the ceiling. If so, how will you know if you can?

    a. No
    b. Yes if it is allowed in the inverter installation manual
    c. Yes if the ceiling is less than five feet
    d. Yes under any circumstances

48. What is the minimum depth of working space that you must have facing a 5kW residential inverter mounted on a wall between the inverter and a wooden wall facing the inverter?

    a. 18 inches
    b. 30 inches
    c. 24 inches
    d. 36 inches

49. In Hawaii, if you have a low-slope rooftop with unlimited space for your PV system, what is the best tilt angle for optimal kWh/kWp/yr production?

    a. Latitude
    b. Flat
    c. Latitude +15
    d. Latitude −15

50. Which of the following is the most important thing you should tell your customer about having a three-degree tilt with a typical PV system?

    a. They should be prepared to have less production due to reflection
    b. They will have more wind loading than a 20-degree tilt
    c. Airflow will be better with a low tilt angle and increased voltage
    d. Expect more soiling losses

51. Of the following, which are the most important things to give a customer with a proposal for a PV system?

    a. String sizing calculations
    b. Size of the system and module datasheets
    c. Expected low temperature
    d. Wire sizes

52. A utility bill has data for May and June and you are asked to give an estimate on a PV system and to size the appropriate system with the data given. Assume every month of the year has similar usage.

MAY	
1427KWH	$214
DEMAND CHARGES	$117
JUNE	
1398KWH	$210
DEMAND CHARGES	$115

If a PV system performance at the location would be 1457kWh/kWp/yr, how much of the bill could you offset and what would be the size of the system?

    a. Offset $3936 with 11.6kW
    b. Offset $2544 with 11.6kW
    c. Offset $656 with 15.2kW
    d. Offset $10,206 with 33kW

53. Your system is making 245,000kWh per year when installed. How much power would you expect the system to make in 10 years with a conservative estimate of 1% per year degradation?

    a. 239,000kWh/yr
    b. 242,000kWh/yr
    c. 237MWh/yr
    d. 222MWh/yr

54. What is the typical process for getting an interconnection agreement for a residential 4kW system?

    a. Send in the utility paperwork and have a bidirectional meter installed before doing any work

    b. Send in utility paperwork during construction

    c. Send in utility paperwork including signed off permit after construction is completed

    d. Send in utility paperwork after utility inspection and meter change

55. How long will it take for a typical PV system to offset its carbon footprint, including the PV inverter and balance of systems?

    a. Less than 6 months

    b. 1 to 2 years

    c. 5 years

    d. 10 years

56. What will you do to determine if you can install a PV system on an industrial metal building in Buffalo, New York?

    a. Verify that the racking system can support the system on the building by reading the installation manual

    b. Hire a structural engineer

    c. Get permission from the Buffalo Chamber of Commerce

    d. Get an interconnection agreement

57. Your customer is an ecologically conscious consumer and you can tell that she is interested in doing the right thing. What should you emphasize when generating a proposal for her?

    a. Carbon offsets

    b. Nitrogen fixing

    c. How the PPA can sell tax credits on Wall Street

    d. How you calculate the string length

58. You are installing a system that should make 1360kWh for every kW installed. The system has had some unforeseen losses since the modules are in a place where it is very dusty and the customer is not cleaning the modules, which leads to 12% of extra losses. If the system makes 10,100kWh per year after the losses from soiling, how large is the PV system?

   a. 4.8kW
   b. 8.4kW
   c. 9.5kW
   d. 10.6kW

59. With southeast- and south-facing arrays of thirteen 240W modules each, what is the best inverter combination for maximum production? All inverters listed have a single MPP.

   a. Two 3kW inverters
   b. One 7kW inverter
   c. Three 2.5kW inverters
   d. One 2kW inverter and one 8kW inverter

60. With a 225A busbar and a 175A main breaker, what is the largest inverter you can install using the 120% rule on a 240V service?

   a. 22kW
   b. 14kW
   c. 5,200W
   d. 18.2kW

61. What is the least expensive way to install a 4kW inverter on a house with a 100A busbar and a 100A main breaker?

   a. 120% rule load-side connection
   b. Supply-side connection
   c. Upgrade service panel to 200A
   d. Line side tap

62. Which will have better energy production?

   a. Ground mount
   b. Pole mount
   c. Flush roof mount
   d. Ballasted roof mount

63. You are selling a ground mount at a school and the school is concerned about children getting electrocuted by the array. What should you tell the school?

    a. There will be a fence around the array

    b. Children would get shocked, not electrocuted in a worst-case scenario

    c. The connectors between the PV modules are the locking and latching type and will not allow the children to get shocked

    d. There is more danger from changing a light bulb than from a PV array

64. In a location at 45 degrees latitude, you have a roof that is sloped 30 degrees to the south. If you are going to mount PV on that roof, what would be the best way to mount the PV?

    a. Reverse tilt on north side of roof, so that the PV is tilted 45 degrees to the south

    b. Extra tilt with PV mounted on the south side of roof, so that PV is tilted at 45 degrees to the south

    c. PV tilted on the south side of the roof to 40 degrees south

    d. PV installed with the slope of the roof

65. Who gets the final say in whether you can build on a rooftop that is questionably stable when the weight of a PV system is added?

    a. AHJ

    b. Racking manufacturer

    c. Structural engineer

    d. EPC contractor

66. What would happen if the installation crew used the smallest wire required by the code from an inverter to the point of common coupling, which was 2000 feet away?

    a. The inverter would likely anti-island

    b. The inverter would stay on, but make less

    c. The wire would burn

    d. The overcurrent protection devices would not function

67. With 15% efficient PV, what would be the most PV that you could fit on a rooftop that is 390 square feet?

    a.   3.1kW
    b.   4.2kW
    c.   5.4kW
    d.   6.3kW

68. What is the main reason to know the low temperature for designing a PV system?

    a.   Low temperatures can freeze solar cells
    b.   Low temperatures cause voltage to decrease
    c.   Low temperatures cause current to increase
    d.   Low temperatures cause voltage to increase

69. You sell a 100kW PV system and the customer complains that the system is not performing close to 100kW. You go to see him and the system is operating on a summer day and the cell temperature is 65C. The irradiance in the plane of the array is 900W per square meter and other derating factors are determined to be 0.77. What would you expect the output of the inverter to be if the temperature coefficient for power is –0.5%/C?

    a.   55kW
    b.   70kW
    c.   80kW
    d.   33kW

70. With a 3kW inverter that can have five to ten 175W modules in series, you would like to install 18 modules all facing the same direction. Five of the modules will be shaded for an average of two hours per day. The inverter has a single MPP input. What would be the best configuration of the following?

    a.   Two PV source circuits of nine modules in series
    b.   One PV source circuit of eight and another of ten
    c.   One source circuit of five, another of six and another of seven
    d.   Three PV source circuits of six

70-question practice exam with detailed explanations

Chapter 8

1. A south-facing roof has measurements of 91 feet north to south and 154 feet east to west. What is a rough estimate of how much PV you can fit on the roof?

   a. **140kW**
   b. 15kW
   c. 500kW
   d. 90kW

A quick way to estimate how much PV will fit on a rooftop is to use 10W per square foot. This estimate is an underestimate and can be used to account for extra roof space for walkways, fire department offsets, skylights, air conditioners, inter-row spacing, etc.

$$91ft \times 154ft = 14,014 \text{ square feet}$$
$$14,014 \text{ square feet} \times 10 \text{ watts per square feet} = 140,140W$$
$$140,140W / 1000 = \textbf{140kW}$$

Estimates like this can be very useful to get us in the ballpark and we can later make a more accurate estimate by doing a layout using CAD software.

2. Which are the best tools for measuring solar access?

   a. Tape measure, pyranometer and transit
   b. Pyrheliometer, temperature sensor, calculator
   c. **Sunpath calculator, compass**
   d. IV curve tester, compass

Solar access is determined by knowing how much open sky there is in relationship to shading objects, and taking into consideration where the sun is inside the solar window.

A sunpath calculator, which usually includes a compass, is often used to determine solar access. The two most popular sunpath calculators are the Solar Pathfinder and the Solmetric Suneye (the Suneye is no longer in production, but still popular).

A pyranometer is an irradiance-measuring (brightness-measuring) device. It would be inconvenient to have to use a pyranometer to measure solar access, since we would have to measure with our pyranometer for years to get good data.

A pyrheliometer is used to measure direct beam radiation. It is useful to have direct beam radiation data when planning a concentrating PV system, which is rare.

Other answers could be argued for this question, but the best answer includes a sunpath calculator.

3. Which of the following is a PBI (performance based incentive)?

    a. Investment tax credit (ITC)
    b. Depreciation
    c. Feed-in-tariff (FIT)
    d. Time of use (TOU)

A performance-based incentive is one that pays you for solar energy produced, not for how much you spent on your PV system, or how big your PV system is. One way to look at this is that if you installed your PV system under a tree, would you get the same incentive? If the answer is yes, then it would not be an incentive based on performance.

A FIT is a program, which was invented in Germany, which pays a system owner typically a fixed price per kWh for usually 20 years. It is essentially a contract that the utility is required to pay due to a law. FIT programs are most popular in Europe and the Ontario Province of Canada. Since the FIT incentive is based on energy production, it is a PBI.

An ITC is based on a percentage of the price of a PV system and a person will get a credit for paying taxes based on the ITC percentage of the price of the system.

Depreciation is also based on the cost of the system. A business may depreciate the value of a PV system over time and there are different types of depreciation.

ITC and depreciation would work just as well for a PV system built under a tree as for a PV system with good solar access.

Time of use is a utility rate schedule and not as good an example of a PBI as a FIT.

4. Southeast- and southwest-facing roofs are going to have solar on them. Each rooftop will fit seven 250W modules on it. What would be the best inverter combination for this project? Assume that the given PV source circuit sizes will adequately configure with the inverter.

   a. 1 PV source circuit of 14 modules in series on a 3.8kW inverter with 2 MPPs
   b. 2 PV source circuits of 7 modules each on a 3.5kW inverter with a single MPP
   c. 2 separate 2kW inverters
   d. **A single 3kW inverter with 2 MPPs**

Analyze answer A:

We would not put modules on a single PV source circuit with different orientations. This would be the worst solution. When there are modules with different orientations in the same PV source circuit (string), then whatever module is getting the least amount of light in the plane of the array would determine the current of the entire string.

Analyze answer B:

Having two source circuits on a single MPP with different orientations for each source circuit is acceptable and has been done many times by many solar engineers. In this case with two strings and different orientations, the currents of the different strings would add and the voltage would probably be close from one PV source circuit to the other; however, there is a better answer.

Analyze answer C:

Two separate inverters would operate as well as any solution. The only arguments against two separate inverters is that there would be more costs associated with installing two inverters, including the inverter cost, installation of two separate inverters, extra ac system components and then replacement and maintenance of separate inverters as time goes on.

Analyze answer D:

A single 3kW inverter with two separate MPPs would be a great solution for this project. Even though there is 3.5kW of PV, this system wouldn't be able to produce 3kW unless it was on the top of Mount Everest (exceptionally cold and bright). One reason this project would never clip (produce 3kW when it could produce more) is that the arrays are facing different directions. When the southeast array is directly facing the sun, the southwest array would be 90 degrees azimuth off from directly facing the sun. This inverter could track each array differently. If there were a shadow from a bald eagle sitting on one module, the bypass diodes would kick in and just bypass one module without affecting anything else. Additionally, with option D there is only one inverter, so there will be only the costs of installing one inverter and also there will be only the costs of replacing one inverter at the inverter's end of life.

5. On a residential two-story composition asphalt shingle roof, you discovered that you can fit more modules by orienting them in landscape mode and you are using a standard rail system. The rail instructions say that the rail must be mounted perpendicular to the rafters, which go from the ridge to the eaves. What must be taken into consideration before mounting the modules?

    a. Do the manufacturer's instructions let you mount the modules by connecting to the short edges?

    b. Are the rafters 24" on center?

    c. Do the rails have splice kits that allow for expansion?

    d. Use three rails per module.

As with anything, it is important that the equipment was designed to do the job we are contemplating. The rails and modules were tested according to the manufacturer's directions and we must always follow the manufacturer's instructions when installing the project. Some modules are not designed to be mounted landscape with the attachments on the short edges (and some are). Problems with mounting modules on the short edges can be instability due to wind blowing on the modules and a longer distance between attachments. Also, wind will be more likely to cause vibrations and noises. Perhaps snow loading was also not tested with the attachments on the short edges of the modules.

Rafters 24" on center are common, but there are other dimensions that may work with the installation. Splice kits that can be used for expansion are often used when rails have to go a longer length, especially in a location where there is a larger difference between hot and cold temperatures. If a module requires three rails per module, you must comply; however, modules that are not allowed to be mounted with clamps on the short edges are still not going to be allowed to be mounted with clamps on the short edges, no matter how many rails there are.

6. In Calgary, Alberta, there is a tendency for snow, wind and rain. A big-box retail store would like to have as much solar PV as possible installed on the roof. In order to not waste time on a job that may never happen, what would be the best course of action?

   a. Avoid snowdrift by mounting the system with an increased tilt angle.
   b. **Hire a structural engineer to determine whether the roof can take the weight of the PV system.**
   c. Design a ballasted system with wind deflectors to avoid extra ballast weight.
   d. Design a penetrating system.

Especially in locations with snow loads, it is prudent to have a structural engineer involved with the project early, in order to determine if and how PV may be mounted on the building. Too often, solar salespeople spend much time and resources developing a project, only to find out later that the project is not feasible due to the building needing expensive reinforcements in order to support the PV system and the snow.

In many places with snow, the roof has to be shoveled off once in a great while, when the snow is deep and if there is rain on top of the snow. It would be difficult if not impossible to remove the snow that is in a solar array.

Additionally, there have been jobs in the snow country where the structural engineer required that the PV be mounted flat, due to snow drift. If the modules are flat, the snow can blow away; if the modules are tilted, they can trap heavy snow up on the rooftop.

Wind deflectors can help lower the ballast weight required to keep a ballasted system on the roof, but still with any system in snow country, it is good to find out early from a structural engineer if any solar PV may be mounted on the roof.

Penetrating systems may work better in snow country because of not needing the extra weight of ballast, but there are also other issues that may prevent the installation of the system that a structural engineer can determine.

It is interesting to note that there was once a company called Solyndra that designed a non-penetrating system that was light and could withstand winds of up to 130 miles per hour. Unfortunately, the company had financial troubles due to increased pressures from unprecedented low module prices. There are now rooftops that do not have a solar option, since Solyndra went out of business.

7. Who will approve seismic reinforcements on a commercial rooftop ballasted PV system job in a known earthquake fault zone?

    a. Racking manufacturer
    b. Module manufacturer
    c. Inverter manufacturer
    d. **Building department**

The final approval for everything is going to be the building department. This question has multiple answers that could be argued to be important considerations, but only one best answer. Everything must go through the building department, which is also known as the AHJ or authority having jurisdiction. The AHJ always has the final say. There are other forms of the AHJ, such as the utility, but when we usually talk about the AHJ, we are referring to the building department. The building department also has the authority to interpret the NEC and to add whatever they deem necessary to the requirements and regulations for the installation of PV systems.

8. You are selling and designing a 100kW array that has a single inverter that has seven modules shaded about two hours per day. What is the best way to get maximum production out of this system? The PV source circuits are ten modules in series.

    a. Have each of the seven different shaded modules on a different PV source circuit
    b. **Have all of the modules on the same PV source circuit**
    c. Have three shaded modules on one PV source circuit and four shaded modules on another PV source circuit
    d. Have one shaded module on one PV source circuit, two shaded

modules on another PV source circuit and four shaded modules on
another PV source circuit

The best answer is B: to have all of the shaded modules on the same PV source
circuit, so that the shading only affects a single PV source circuit. With the different PV source circuits and assuming that the 100kW system has a single MPP,
then the shading will only affect a single string. If the shaded modules were on
seven separate strings, the IV curves of seven strings, which would include 70
modules, would be working off of the MPP. In the other examples, it would also
affect more than a single string of modules.

9. If you are canvassing locations and you do not go to the house where you
   would like to install PV, what would be the best way to get information
   about the feasibility of the job?

   a. NREL Redbook data
   b. Satellite data
   c. Weather data
   d. Homeowner opinion

Many solar companies make proposals and some even make contracts based
on satellite data. It is still recommended to go to the house and see the quality
of the roof and other information that cannot be properly accessed based on
satellite data.

NREL Redbook data used to be used to access insolation data, but these days we
use more accurate software.

Weather data are also important, but by itself we cannot determine production
and engineering of a system. Just because you know the weather or the insolation does not mean you will know how much PV will fit on a rooftop exposed to
the sun at a good orientation.

Only if the homeowner is a solar PV expert should we be able to invest in the
homeowner opinion, and usually that is not the case.

10. A customer has a time of use (TOU) utility rate schedule. Their off-peak
    rates are $0.10 per kWh, their partial peak rates are $0.12 per kWh and
    their peak rates are $0.15 per kWh. The customer's total energy bill for the
    month of August was $2000 and they used 3000kWh off-peak and spent
    $200 on partial peak energy. How much peak energy did they use?

    a.   1.2MWh
    b.   10,000kW
    c.   **10MWh**
    d.   5,000kWh

3000kWh off-peak $\times$ $0.10 = $300
$2000 bill – ($300 off-peak + $200 partial peak) = $1500 peak use
$1500 peak / $0.15 per kWh = 10,000kWh peak use

There are many ways to go about solving this problem and I chose to solve first for dollars and then convert to kWh.

Be careful of units. I recommend looking at all of the answers before solving the problem and crossing off every answer that is given in the wrong units. This problem asked for energy usage and not power usage, so the 10,000kW answer is obviously wrong.

Practice converting Wh to kWh to MWh back and forth. One easy way to remember how to do this is that it is all about three decimal places and we can move the decimal where the comma goes when making long numbers:

1,000,000Wh = 1,000kWh = 1MWh

11. What is the social benefit of having a MW of PV on your rooftop?

    a. CO_2 reduction
    b.   Saving oil
    c.   **Creating green jobs**
    d.   SRECS

The NABCEP PV Technical Sales Exam has categories that have to have questions written for them. I can imagine that it would be difficult to come up with a good exam question on the social benefits of having solar, so look out for some good common sense questions that even a homeowner should be able to answer.

Creating green jobs is the most social answer.

12. When is the appropriate time to tell a cash customer about the permit costs associated with installing a PV system within their particular building department zone?

    a.   When the permit package is filed
    b.   **At the proposal stage**

    c.   After the contract is signed

    d.   At the filing of the interconnection agreement

The best and most ethical time to tell the customer about all of the fees associated with going solar is in the beginning, so that they are not surprised about added fees. If we tell the customer about added fees after the contract is signed, which would be the case with three out of four of these answers, then the customer would likely be upset.

The order of these answers is first the proposal is presented to the customer, then the contract. After the contract is signed engineering is done and then a permit package is submitted to the building department. After the permit package is approved, then the PV system is built. After the system is built, then the system is inspected and the inspector signs off the permit. A copy of the permit is then submitted as part of a package for an interconnection agreement.

13. If you were using frameless PV modules (laminates) rather than framed modules at a low tilt angle, what would be a benefit in the context of less derating for production?

    a.   Lower price of modules

    b.   Manufacturer's production tolerance

    c.   Module mismatch

    d.   **Soiling**

Modules with frames at a low tilt angle have a tendency to have higher deratings due to damming of dust and dirt at the lower part of the module above the frame. Sometimes half of a row of solar cells can be shaded with a very low tilt and a framed module in a location where it does not rain often.

At first we might try to consider this effect a form of module mismatch; however, this is not a result of the manufacturing of the module, it is a result of soiling.

14. In a location that would get production of 1.3MWh per year for every kW of PV installed, how large of a system would we need if we were to offset all of our energy and used 14,000kWh per year?

    a.   7kW

    b.   8.1kW

    c.   9.5kW

    d.   **10.8kW**

Since everything in this question is given in kW or kWh, let us start by converting 1.3MWh to 1300kWh.

Many of us in the industry would say that this location will get 1300kWh/kWp/yr, which means that if we installed a kW of PV we would get 1300kWh per year of production. The p from kWp is for kW peak or peak sun, which is STC or 1000W per square meter of irradiance, which is how all PV is rated.

If we need to produce 14,000kWh per year and each kW will produce 1300kWh per year, we can divide:

$$14,000\text{kWh per year} / 1300\text{kWh/kWp/yr} = 10.8\text{kW}$$

We would need to have 10.8kW of PV and we can double-check our answer by multiplying:

$$10.8\text{kW} \times 1300\text{kWh/kWp/yr} = 14,000\text{kW}$$

15. A customer has a bill for $5000 for a month of energy with a tiered rate schedule. There are equal charges for three different tiers for two-thirds of the bill and there is a demand charge for one-third of the bill. About how much can they save by getting a PV system?

   a. $5000
   b. $3400
   c. $4000
   d. $1200

It is generally accepted that solar alone without batteries cannot help much with demand charges. Demand charges are charges for power and not energy. Typically the most power a customer will use in a month will cause a demand charge. The problem with solar offsetting demand charges is that there will be demand when there is no sunlight or when there are poor sunlight conditions.

Many unknowledgeable or unethical solar salespeople have promised to offset 100% of a customer's bill when there are demand charges.

Since in this case the demand charges were about one-third of the bill, then we were only able to offset about two-thirds of the $5000 bill. We may reduce the demand portion a very small amount, but not much. Two-thirds of $5000 is $3333 and we can round up to $3400, since we may lower demand slightly occasionally.

Demand charges are most common with commercial electric bills.

The solar can offset all of the energy for the tiered rates, but usually none to very little of the bill for demand charges.

16. What is the most important factor in a good long-term return on invest-ment for a FIT project?

    a. **kWh per year**
    b. kWh/square meter
    c. Tax equity
    d. Information technology

FIT is a feed-in-tariff and is a performance-based incentive (PBI). The more energy a FIT project makes, the more return on investment the project makes. When a FIT project is designed, the big factor is how many kWh per year the project can make. Typically a FIT will get a fixed amount of money for each kWh for a period of 20 years. The FIT was first introduced in Germany, and is most popular in North America in the Ontario Province of Canada.

17. At what height must a worker use fall protection?

    a. 4 feet
    b. 8 feet
    c. 12 feet
    d. **6 feet**

This should be the easiest question on the exam and is typically on every con-struction-related exam. In the USA fall protection is required at heights above 6 feet. This is why many mock roofs at solar schools are 5'11".

18. Who can open an electrical box according to OSHA?

    a. NABCEP Certified person
    b. **Qualified person**
    c. NABCEP Entry Level Certificate holder
    d. UL Certified Installer

According to the OSHA website, a qualified person is "One who has received training in and has demonstrated skills and knowledge in the construction and operation of electric equipment and installations and the hazards involved."

Someone who is NABCEP Certified would likely be qualified; however, the OSHA inspector may or may not consider him or her qualified. OSHA does not mention NABCEP by name.

The same goes for a UL Certified Installer. OSHA does not decide whether NABCEP or UL Installer Certifications are more desirable.

The NEC also requires that a qualified person install all equipment and associated wiring and interconnections for PV systems.

19. What is the official term that we use to call all different forms of materials that we use to keep workers safe?

    a. Safety precaution materials
    b. Safety protective equipment
    c. Personnel protective equipment
    d. Personal protective equipment

The things that we use to protect ourselves on the jobsite are officially called personal protective equipment (PPE). This includes fall protection equipment, earplugs, eye protection, hard hats, etc.

20. What determines the maximum size of a supply-side inverter connection?

    a. 120% busbar – 125% of the inverter current
    b. 120% busbar – solar breaker
    c. The rating of the service
    d. Loads

A supply-side connection is between the meter and the main breaker (main service disconnect). When we are on the supply side of the main breaker, the main breaker is already protecting the busbar from the utility and the utility has a lot of current compared to the inverter, which makes the current from the inverter almost insignificant as far as protecting the busbar is concerned. What limits the size of the inverter on a supply-side connection is the rating of the service. Service entrance conductors are the conductors connected to the meter that the utility "serves" the customer. It would be very difficult to imagine a situation where someone would want to have an inverter that would put out more current than the service entrance conductors can handle. Nobody uses that much energy, or he or she would have a larger service.

21. What size overcurrent protection device would you use for the ac output of a 7kW inverter on a single-family house?

    a. 25A
    b. 40A

c. 30A

d. 45A

To size a breaker, you have to find out the current of the inverter and multiply by a correction factor of 1.25 and then round up to the next common overcurrent protection device size.

To calculate the current of the device we will use the equation:

$$watts = volts \times amps$$

Then solve for amps:

$$amps = watts / volts$$

Since the volts for a single-family dwelling are 240V and the power in watts of a 7kW inverter is 7000W, then:

$$amps = 7000W / 240V = 29.2A$$

Now we round up 29.2A to the next common breaker size.

Common overcurrent protection device sizes above 15A increase by increments of 5A until 50A, and then increase by 10A to 110A (15, 20, 25, 30, 35, 40, 45, 50, 60, 70, 80, 90, 100, 110).

Some breakers, such as a 25A breaker, are difficult to find and in the field inspectors have been known to let people use a 30A breaker in place of a 25A breaker.

29.2A rounds up to 30A, so we would use a 30A breaker for a 7kW inverter on a house.

22. What is the voltage of a typical small commercial service?

    a. 120/240V wye

    b. 240/208V

    c. **120/208V**

    d. 240/208V

Small commercial voltage is 120/208V, which is three-phase. One way you can determine if something is three-phase is to take the small number and multiply by the square root of 3, or take the big number and divide by the square root of 3. The square root of 3 is about 1.73

$$120V \times 1.73 = 208V$$

120/208V is three-phase small commercial voltage.

As a side note, when we are looking at larger commercial buildings, we will see 277/480V, which is also related by the square root of 3.

23. You notice that the inverter says it is making more energy in the winter than it ever did in the summer. What is most likely happening?

    a. **Cold and wind increases production**
    b. Inverter display broken
    c. Array was broken and wind blew it back in place
    d. Azimuth must have been more favorable for winter production

The most likely scenario is that the cold temperatures and the wind are increasing voltage and production. The wind will blow the heat caused by the sun away from the module. Also, on a cold day there is low humidity and that will lead to brighter light.

If the temperature coefficient of power is about −0.48%/C and the low temperature is −40C and with a high windspeed, the PV could be operating at −35C. We can compare this to the summertime high temperature of 40C and extrapolate another 30C increase in cell temperature with no wind in the summer for a cell temperature of 70C, which is 105C warmer than −35C.

The difference in power due to temperature would be:

$$105C \times 0.48\%/C = 50\% \text{ increase in power due to temperatures}$$

Another factor in the increased performance on this unusual winter day could be reflections from snow on the ground increasing irradiance.

In general, PV works best in the summer because of the increased sunlight; however, conditions can be just right to have extra amounts of power in the winter.

24. An inverter can have a top voltage of 500V and the PV being used has an open circuit voltage of 36.8V. With a low temperature of −30C and a typical voltage temperature coefficient, what would be the maximum number of modules in series?

    a. 14
    b. 8

c. 9

d. 11

The big question here is: what is a typical temperature coefficient of Voc? Usually PV will have an increase of about one-third of a percent of voltage for every degree C decrease from standard test conditions (STC).

First the difference in temperature from STC, which is 25C, and the cold design temperature is:

$$-30C - 25C = -55C$$

Now we multiply the difference in temperature by the estimated temperature coefficient:

$$-55C \times -0.33\%/C = 18\% \text{ increase in voltage}$$

Turn the percentage into a decimal:

$$18\% / 100\% = 0.18$$

Add 1 to make it an increase in voltage and a temperature correction factor:

$$1 + 0.18 = 1.18$$

Multiply the Voc by the temperature correction factor to get the cold temperature Voc for the module:

$$1.18 \times 36.8V = 43.4V \text{ when cold}$$

Now we determine how many times the cold temperature Voc will go into the inverter maximum input voltage:

$$500V / 43.4V = 11.52$$

Since we cannot have half of a module in series and since we cannot go over voltage by rounding up, we will round down to **11 modules in series maximum**.

25. You are installing a large commercial rooftop PV system in stages. The first stage is 500kW, and at the end of year 1 the system makes 704MWh of energy, preventing 528 tons of carbon dioxide from going into the atmosphere. The next stage of the project is to install 258kW with the same expected performance. At the end of the year after stages 1 and 2 are installed, how much carbon dioxide will the entire system prevent from going into the atmosphere?

a. 754 tons
b. 595 tons
c. 701 tons
d. 800 tons

If a 500kW system will offset 528 tons of carbon dioxide per year, then we can calculate how many tons of carbon dioxide each kW of PV will offset in this particular location:

528 tons / 500kW = 1.056 tons per kW

The new system will have two stages added together; stage 1 is 500kW and stage 2 is 258kW:

500kW + 258kW = 758kW total

Now if each kW offsets 1.056 tons of carbon dioxide and we have a total of 758kW of PV, then:

758kW × 1.056 = **800 tons of carbon dioxide per year**

If we had PV in a different location or a different orientation, we could determine how many tons of carbon dioxide 1MWh of solar produced energy would prevent from going into the atmosphere. We could do this calculation by calculating tons per MWh:

528 tons / 704MWh = 0.75 tons per MWh

This would be a conversion factor that would work under different circumstances. The reason we did not convert with this conversion factor was because it was simpler and required fewer calculations to calculate for tons per kW. The reason tons per kW will not always work is that a kW will make different amounts of energy in different places.

26. An optimally placed PV system in a location in the Nevada Desert makes 1800kWh/kWp/yr. If there were another system being installed in the same way a mile away that was 30kW, then how many MWh would you expect the system to make in 20 years?

   a. 76
   b. 54
   c. 1080
   d. 1180

There are many ways to calculate this equation. First, we will convert 1800kWh/kWp/yr to MWh/kWp/year, since we want our answer to be in MWh.

1800kWh/kWp/yr / 1000 = 1.8 MWh/kWp/yr

Next, we will determine how many MWh 30kW will produce in a year:

30kW × 1.8MWh/kWp/yr = 54MWh per year

The system will produce 54MWh per year, which we just need to multiply by 20 years:

54MWh per year × 20 years = 1080MWh

As a note: when we see kWp, that means kW of PV at STC, which is 25C, 1000W per square meter and 1.5 air mass, which is how all PV is tested and sold.

27. A typical 3kW utility interactive PV system produces 8% less per year the second year than it did the first year. What is the most likely problem of the following? You checked the system and it appears to be working properly.

    a. Blown PV source circuit fuse in combiner
    b. Weather fluctuations
    c. Module performance degradation
    d. Magnetic declination

A typical 3kW system will have one or two PV source circuits and if there were a blown fuse the affected PV source circuit would not work at all. Losing one source circuit out of one or two would be a 100% or a 50% loss of production.

Module performance degradation on a typical 3kW PV system would be less than 1% per year and would not lead to an 8% loss of production.

Magnetic declination would have nothing to do with year-to-year performance.

Weather patterns can produce minor differences in production, such as a drought, the end of a drought or a very rainy or cloudy year.

28. What is the purpose of calculating magnetic declination?

    a. To determine the longitude
    b. To determine the latitude
    c. To determine the tilt angle
    d. To determine azimuth

The magnetic north pole is in a different place than the geographic north pole, which is the axis of the Earth as it spins. Using a magnetic compass will only give us information about where the magnetic north pole is. We will have to add or subtract to determine where true north or south is located. True azimuth will tell us where the sunpaths are in the sky. Almost all information is given to us in true azimuth, except for a magnetic compass and airport runways, which are also lined up with magnetic azimuth.

29. What would be the best, most cost-effective way to interconnect a 4kW STC PV system on a 120/240V residential service? The house has a typical 100A service with a 100A busbar and a 100A main breaker on the top of the busbar.

   a. With a 4kW inverter and a supply-side connection with a 25A inverter overcurrent protection device.

   b. With a 3.8kW inverter on a 20A circuit breaker with a load-side connection.

   c. With a 3.8kW inverter on a supply side connection with a 20A overcurrent protection device.

   d. With a 4kW inverter on a 20A circuit breaker with a load-side connection.

For sizing the proper inverter for 4kW of PV, 3.8kW of inverter is a ratio of:

$$4kW:3.8kW = 1.05:1 \text{ ratio}$$

If we had 4kW of PV on a 3kW inverter that would be a 1.33:1 PV to inverter ratio and in some cases the PV may have the ability to produce more than the inverter could put out, which is not the end of the world and it is something we call clipping power. Many larger PV systems clip power on a regular basis.

Many good solar engineers will design PV systems with a 1.2:1 PV to inverter ratio or more. There is no scenario where 4kW of PV would be too much for a 3.8kW inverter.

A load-side connection is less expensive and easier to perform than a supply-side connection. We will now apply the 120% rule and see what is the largest inverter we can place on a supply-side connection with a 100A busbar and a 100A main breaker.

The 120% rule states that 125% of the inverter current plus the main breaker may not exceed 120% of the busbar rating. Also, when we do exceed the busbar rating, we have to make sure that the solar breaker is on the opposite side of the busbar from the main breaker.

To make it simpler, let us look at the 120% rule in a different way:

$$\text{busbar} \times 1.2 \geq \text{main} + (\text{inverter current} \times 1.25)$$
$$(\text{busbar} \times 1.2) - \text{main} \geq \text{inverter current} \times 1.25$$

In the 2011 NEC and earlier we used the solar breaker size in the calculations. In the 2014 NEC we use 125% of the inverter current in the calculations for the 120% rule. It is more confusing with the 2014 Code, since we are talking about 125% of the inverter current for the 120% rule. The benefit is that when we size a breaker, we take 125% of the inverter current and then round-up to the next largest common breaker size. With the 2014 Code we are not penalized for the rounding up to the next common larger circuit breaker.

Let us plug in some numbers:

- busbar = 100A
- main breaker = 100A

We can first solve for 125% of inverter current:

$$125\% \text{ of inverter current} \leq (\text{busbar} \times 1.2) - \text{main breaker}$$
$$125\% \text{ of inverter current} \leq (100A \times 1.2) - 100A$$
$$125\% \text{ of inverter current} \leq 120A - 100A$$
$$125\% \text{ of inverter current} \leq 20A$$

So now we know that:

$$\text{inverter current} \times 1.25 \leq 20A$$

So with simple algebra:

$$\text{inverter current} \leq 20A/1.25$$
$$\text{inverter current} \leq 16A$$

To find out what the power of a 16A inverter is at 240V, we will calculate:

$$\text{watts} = \text{volts} \times \text{amps}$$
$$\text{watts} = 240V \times 16A$$
$$\text{watts} = 3840W$$

The largest inverter that we can use in this situation on a load-side connection is 3840W, which is a 3.8kW inverter.

It is not a coincidence that 3.8kW inverters are more popular than 4kW inverters. They make them this way for a purpose.

If we were to use a 4kW inverter, we would not be able to do a load-side connection and would be stuck doing a more expensive supply-side connection. Some of the reasons that a supply-side connection is more expensive are:

- A supply-side connection requires turning off the power to the service entrance conductors between the meter and the main breaker. This usually involves pulling the utility meter. In some places the utility will let a solar installer pull the meter and in other places the utility has to be there to pull the meter.
- A supply-side connection requires connecting to large service entrance conductors and installing equipment that can withstand high currents coming from the utility. We usually use service rated equipment for supply-side connections.
- Load-side connections are very easy. We switch off the main breaker for a short time while we pop in a solar breaker and it is done!

A 3.8kW inverter will use a 20A overcurrent protection device.

To calculate breaker size, we first calculate inverter current:

$$\text{watts} = \text{volts} \times \text{amps}$$
$$\text{amps} = \text{watts} / \text{volts}$$
$$\text{amps} = 3800W / 240V$$
$$\text{amps} = 15.8A$$

Then, to find the overcurrent protection device size, we multiply by 1.25 and round-up to the next common overcurrent protection device size:

$$15.8A \times 1.25 = 19.75A$$

Round-up to 20A overcurrent protection device size.

The correct answer here is B, since a 3.8kW inverter is most cost-effective as it can be connected with a load-side connection and a 20A overcurrent protection device.

30. Which of the following is the largest inverter you can connect on a load-side connection with a 225A busbar and a 200A main breaker on a typical residential main service panel?

    a. 7kW

    b. 8kW

    c. 12kW

    d. 15kW

This is another 120% rule question and we will use much of the same calculations that we used in practice question number 29.

$$\text{busbar} \times 1.2 \geq \text{main} + (\text{inverter current} \times 1.25)$$
$$(\text{busbar} \times 1.2) - \text{main} \geq \text{inverter current} \times 1.25$$
$$\text{inverter current} \leq ((\text{busbar} \times 1.2) - \text{main}) / 1.25$$

Also, we can make a larger formula and solve for inverter power since inverter current multiplied by inverter ac residential voltage (grid voltage) is equal to inverter power:

$$\text{max inverter power} = 240V \times (((\text{busbar} \times 1.2) - \text{main}) / 1.25)$$

Now we can just plug everything into the formula:

$$\text{max inverter power} = 240V \times (((225A \times 1.2) - 200A) / 1.25)$$
$$\text{max inverter power} = 240V \times (((270A) - 200A) / 1.25)$$
$$\text{max inverter power} = 240V \times (70A / 1.25)$$
$$\text{max inverter power} = 240V \times 56A$$
$$\text{max inverter power} = 13,440W$$

In this case, of the possible answers the largest inverter that we can use on a load-side connection is going to be a 12kW inverter.

31. In a location with an insolation of 5.4 peak sun hours and a system with a derating factor of 0.77, how many kWh per year would a 10kW system produce?

a. **15,180kWh per year**

b. 12,000kWh/kWp/yr

c. 1400kWh per year

d. 3750kWh

If we were in a location where the solar resource was 5.4 peak sun hours (PSH) of insolation and the derating factor was 0.77, then to get our ac kWh per day for a kW of PV, we would multiply:

$$5.4PSH \times 0.77 = 4.16kWh \text{ ac per day}$$

and if we wanted to calculate how many kWh per year a kW of PV would produce:

$$4.16kWh \times 365 = 1518kWh/kWp/yr$$

We often use the metric of kWh/kWp/yr to describe how many kWh a kW of PV will make in a year. This is very convenient when comparing locations or when determining how much a system will make by multiplying the kWh/kWp/yr by the system size. Also, we can go backwards; if we know how many kWh we want our system to produce, we can divide by kWh/kWp/yr to get the system size in kW.

We then multiply the kWh/kWp/yr by the system size of 10kW to get our answer:

$$1518kWh/kWp/yr \times 10kW \text{ system size} = 15,180kWh \text{ per year}$$

32. A residential customer is buying a net-metered PV system and they are comparing the system to an investment in the stock market. What benefit does an investment in a PV system have that an investment in green energy stocks does not have?

a. You are taxed on income from a PV system, but not at the capital gains rate

b. The taxes from owning a PV system are lower than the taxes from owning green energy stocks

c. A PV system is greener than green energy stocks

d. **You are not taxed on the money saved by a PV system, but you are taxed on income from the stock market**

For a net-metered PV system, there are no income taxes on the money that you save in energy. In many places there are no taxes from owning a PV system, but it depends on the state and local laws. Most residential PV systems installed in the USA are exempt from an increase in property taxes. As with many exam questions, there are gray areas where there could be multiple right answers and it is your job to pick the answer that the people writing the exam want you to pick. If you disagree with an exam question, you have the right to fill out a form during the exam and complain. There could be an argument for answer B, but there are also good arguments against answer B.

For answer C, it depends on which green energy stocks and how the PV system is installed. There is also a gray area here.

Answer D is the best answer and is something that all solar salespeople tell their cash customers. In tax language, the internal rate of return (IRR) of a PV system will include the fact that the PV system benefits are saving money and are not taxable. If you sell your stock and make a profit, you will then have to pay taxes on your investment. Answer D is the most obvious right choice.

Be prepared to see exam questions with what you think are multiple right answers, and try to think about what the exam writers want you to answer with.

33. You are selling a PV system on a house with a preexisting generator and the customer asks you where you will connect the PV system. He asks if the typical grid-tied PV system will work and feed back to the generator instead of the grid when the power is out. Where will you connect the PV system and will it work with the generator?

    a. Connect the PV system inverter on the load side of the generator automatic transfer switch (ATS) and the inverter will work with the PV system when the grid is down.

    b. Connect the inverter on the supply side of the generator ATS and the inverter will work with the generator when the grid is down.

    c. **Connect the inverter on the supply side of the generator ATS and the inverter will not work with the generator when the grid is down.**

    d. Connect the inverter on the load side of the generator ATS and the inverter will not work when the grid is down.

A utility interactive inverter will not work with the power coming from a typical generator. If the grid goes down and the inverter sees the generator, when the inverter is placed on the load side of the generator ATS, the inverter will try to turn on, but the sine waves, voltage and frequency will not be within the specifications of the inverter and the inverter will not turn on. It would not be healthy for the inverter to constantly see the generator fluctuating sine waves and it is recommended to have the interactive PV system connected to the supply side of the generator ATS.

Inverter manufacturers warn against connecting an interactive inverter to a generator, and if by chance the generator made such clean power that the inverter could turn on for a small time, it could damage the generator by feeding power to the generator.

Interactive inverters should be connected to the utility supply side of an inverter ATS.

34. What is the best application for MACRS?

    a. Putting PV on your own house
    b. **Putting PV on your own business**
    c. Putting PV on your non-profit
    d. Putting PV and wind on your own house

MACRS depreciation is modified accelerated cost recovery system. With this type of depreciation and with all types of depreciation, you can only take advantage of it if it is for a business that pays taxes. If you own your PV on your house, you cannot take advantage of depreciation. If someone else owns PV on your roof and sells you the energy in the form of a power purchase agreement, then they can take advantage of depreciation.

A non-profit does not pay taxes and also cannot take advantage of depreciation.

MACRS depreciation is only beneficial when a business owns the PV system.

35. If an 11kW PV system will make 14,895kWh per year, then how much energy will a similar PV system make in the same location if it consists of 48 PV modules with the following specifications? Voc = 37.5V, Vmp = 30.4V, Imp = 8.56A and Isc = 9.12A.

    a. 2010kWh per year
    b. 32,800kWh per year
    c. 1,693kWh per year
    d. **17MWh per year**

If 11kW of PV will make 14,895kWh per year, then we can figure out how many kWh 1kW will make in a year:

$$14,895\text{kWh per year} / 11\text{kW} = 1354\text{kWh/kWp/yr}$$

Then we need to figure out how much PV we have. Given the data, we can multiply the operating STC voltages and current to get the power of the modules:

$$\text{Imp} \times \text{Vmp} = \text{STC watts}$$
$$8.56\text{A} \times 30.4\text{V} = 260\text{W modules}$$

(Make sure to not multiply Isc or Voc for power, because at Isc or Voc there is no power.)

$$48 \text{ modules} \times 260\text{W modules} = 12.5\text{kW}$$

Now we multiply our system size in kW by our kWh/kWp/yr:

$$12.5\text{kW} \times 1354\text{kWh/kWp/yr} = 16,925\text{kWh}$$

We can round it off to 17,000kWh or 17MWh per year.

36. There is a house with a 100A main utility disconnect outside of the house and the 125A service panel on the inside of the house. If planning to connect a 10kW inverter to the house, what is the best place to connect the inverter? There are 90A loads connected to the service panel, the conductor going from the main utility disconnect to the service panel is rated for 130A and you are avoiding a connection to a feeder.

    a.   Between the main utility disconnect and the service panel

    b.   On the far end of the busbar in the service panel

    c.   **Between the main utility disconnect and the meter**

    d.   On the supply side of the meter

First, we will determine if we can do a load-side connection; the largest inverter that we can put on the load side of the main service disconnect according to the 120% rule can be calculated by:

$$125\text{A busbar} \times 1.2 = 150\text{A}$$
$$150\text{A} - 100\text{A main breaker} = 50\text{A maximum solar breaker}$$
$$50\text{A solar breaker} / 1.25 = 40\text{A maximum inverter current}$$
$$40\text{A inverter} \times 240\text{V} = 9600\text{W} = 9.6\text{kW maximum load-side connection}$$

Since we have a 10kW inverter that we would like to connect, we are unable to do a load-side connection according to the information we have. If we had a 9.6kW inverter, that would have been perfect!

There is another new rule in the 2014 NEC that says we can do a load-side connection as long as all of the breakers (solar and load) do not add up to more than the rating of the busbar, as long as we add a sign that says that when we add up all of the breakers they cannot total more than the rating of the busbar. In this case the breakers already on the busbar are 90A and we want to add a 10kW inverter. The size of the breaker for the 10kW inverter would be:

$$10,000W / 240V = 41.7A$$

Then we round-up to a 45A breaker.

For a load center (service panel) with 90A of loads, if we add a 45A solar breaker we would exceed the busbar rating when we sum up the solar and load breakers:

$$45A + 90A = 135A$$

since 135A is greater than the 125A busbar.

I like to call this new ability to do a load-side connection the "sum rule" since we are summing up the solar and load breakers.

Hooking up to the supply side of the meter is not so smart, since we will not get credit for the electricity that we make. Also, if we tamper with the supply side of the meter the utility will usually think we are trying to steal power.

This leaves us with a supply-side connection, which is between the main utility disconnect and the meter. A supply-side connection is never on the load side of any overcurrent protection device and many people mistake a connection on the load side of a main disconnect that is outside of a building going to a service panel in the house as a supply-side connection.

There are other ways that the 2014 NEC allows us to connect to the "feeder" between the main utility disconnect and the service panel.

37. According to the 120% rule, which is the largest inverter of the following that you can fit onto a 200A busbar with a 200A main breaker on a 240V ac inverter?

a. 5kW
b. 6000W
c. **7000W**
d. 8kW

First, we calculate 120% of the busbar:

$$200A \times 1.2 = 240A$$

Then we subtract the main breaker:

$$240A - 200A = 40A$$

Then we either divide 40A by 1.25 or multiply by 0.8 (same thing)

$$40A / 1.25 = 32A$$
$$40A \times 0.8 = 32A$$

Then we multiply V $\times$ I = W:

$$240V \times 32A = 7680W = 7.68kW$$

The largest inverter we could install would be a 7.68kW inverter and the largest from the given choices would be a 7000W inverter.

Remember that the solar breaker would have to be at the opposite end of the busbar from the main breaker and that there would have to be a label at the breaker saying "solar backfeed breaker do not relocate."

38. If a PV system were losing 1% per year due to module degradation, after 30 years what would be the annual loss due to degradation?

a. 30%
b. 70%
c. 74%
d. **26%**

This is a case of compounding losses and can be calculated by first turning the 1% loss into a 99% keep.

$$100\% - 1\% = 99\%$$

99% in decimal form is 0.99.

We then multiply 0.99 by 0.99 30 times.

Instead of hitting so many calculator keys that we lose track, we can use the scientific calculator keys to help us (using a Casio fx260 calculator).

- Enter: .99
- Enter: X^y
- Enter: 30
- Enter: =

Then you should have the number of how much you keep, which is 0.74, which means we keep 74% of our production for the 30th year.

Next, we figure out how much we lose by subtracting 74% from 100%:

$$100\% - 74\% = 26\%$$

We will lose 26% of our annual production with 1% compounding annual losses for 30 years.

You may have to estimate what typical module degradation is and in reality it can often be anywhere between 1% and 0.25%; however, on a test it is recommended to think conservatively and perhaps guess that degradation may be 1% per year.

39. If the price of electricity goes up 3.5% per year for 20 years, what would be the price of electricity 20 years later if the price at the beginning were $0.16 per kWh?

    a. $0.34 per kWh
    b. $0.27 per kWh
    c. **$0.32 per kWh**
    d. $0.29 per kWh

This is a case of compounding annual growth. First, we turn 3.5% into a decimal:

$$3.5\% / 100\% = 0.035$$

Then we add 1, since we are going to have a percentage increase every year.

$$1 + 0.035 = 1.035$$

Then we will multiply 1.035 by 1.035 20 times, or we can use the X^y button:

$$1.035 \times 1.035 \times 1.035 \times 1.035 \times 1.035 \times 1.035 \times 1.035 \times$$
$$1.035 \times 1.035 \times 1.035 \times 1.035 \times 1.035 \times 1.035 \times 1.035 \times$$
$$1.035 \times 1.035 \times 1.035 \times 1.035 \times 1.035 \times 1.035 = 1.99$$

Or

- Enter: 1.035
- Enter: X^y
- Enter: 20
- Enter: =

And you get 1.99, which means your electricity price went up 99%.

Then you multiply:

$$1.99 \times \$0.16 = \$0.32/kWh$$

If electricity rates go up 3.5% per year for 20 years, and we start with $0.16 per kWh of electricity, after 20 years our electricity would cost $0.32 per kWh.

40. If your PV module specifications were Voc = 37.2V, Vmp = 30.1V, Imp = 8.3A and Isc = 8.87A, and you could expect the PV module cell temperature to reach 65C, what would be the minimum number of modules that you could put in series if the inverter MPPT range was 200V to 480V and the inverter maximum input voltage was 600V?

    a.   7

    b.   8

    c.   9

    d.   10

For the high temperature short string calculations, we need to use the high temperature to figure out how low the module Vmp will get and then determine the minimum number of modules that it will take to keep the inverter above the low MPPT voltage.

Since the high temperature is 65C and the information about the module was gathered at STC, which was a temperature of 25C, then the difference in temperature from STC is:

$$65C - 25C = 40C$$

We can say that our delta T is 40C.

Then we will multiply the delta T by the temperature coefficient of Vmp. The temperature coefficient of Vmp is different than the temperature coefficient of Voc. Typically, the temperature coefficient of Vmp is about –0.48%/C, while the temperature coefficient of Voc is closer to –0.33%/C. Also, the temperature coefficient of power is usually the same as the temperature coefficient of Vmp.

$$40C \times -0.48\%/C = 19.2\%$$

This means we will have a 19.2% loss of Vmp. In order to turn the loss into what we keep, we can subtract the loss from 100%:

$$100\% - 19.2\% = 80.8\% \text{ of our voltage we keep}$$

We can turn the percentage into a decimal:

$$80.8\% / 100\% = 0.808$$

and multiply our voltage-derating factor of 0.808 by our Vmp:

$$0.808 \times 30.1V = 24.3V$$

24.3V is our hot temperature Vmp.

Then we divide our hot temperature Vmp into our low inverter MPPT voltage, which is 200V:

$$200V / 24.3V = 8.2 \text{ modules in series}$$

Since we cannot cut a module into pieces, we either have to round-up or round-down. If we round-down our inverter would go under voltage on a hot day, which we do not want, so to stay within the inverter voltage window we will have to round-up to nine modules in series minimum.

You can see from answering this problem that much of the information we did not use. This is always the case in real life.

41. A building has a time of use (TOU) utility rate schedule. The rates are as follows: 9 a.m. to 1 p.m. is $0.12 per kWh, noon to 7 p.m. is $0.22 per kWh and 7 p.m. to 9 a.m. is $0.10 per kWh. If a person uses an even amount of electricity throughout the day, what is the best way to face the PV modules at latitude 33N in a place without dramatic weather patterns?

   a. Southeast
   b. South
   c. **Southwest**
   d. West

Since the best solar energy performance is with the array facing south and the best monetary performance is for the higher tariffs facing southwest and west, then the best performance for saving money should be in between south and west and the best answer is southwest.

42. A residence has a tiered utility rate schedule. The low tier is $0.15 per kWh and includes the first 250kWh per month; the next tier is $0.19 per kWh and includes the next 350kWh; and the final third tier is $0.36 per kWh and includes all remaining kWh. In January, you have used 1103kWh. What would your energy bill be?

    a. $285
    b. $103
    c. $397
    d. $94

With a tiered rate schedule, we have to fill up the lower tiers before we can fill up the higher tiers. A tiered rate schedule encourages energy conservation, so that people are incentivized to use the cheaper energy from the lower tiers. Tiered rate schedules also encourage people that use a lot of energy to get solar more so than people that are in the lower tiers.

> The value of a PV system for someone offsetting tier 3 energy is $0.36 per kWh and for someone offsetting tiers 1 or 2 would be between $0.15 and $0.19 per kWh.

Tiered rate schedules are most common on the west coast of the USA for residential users. Tiered rate schedules are calculated on a monthly basis and we are assuming net metering.

We are using a total of 1103kWh per month.

It is obvious that the first two tiers will be filled and the remaining energy will be in the third tier.

$$\text{tier 1} + \text{tier 2} = 250\text{kWh} + 350\text{kWh} = 600\text{kWh}$$
$$\text{tier 3} = \text{total kWh} - (\text{tier} + \text{tier})$$
$$\text{tier 3} = 1103\text{kWh} - 600\text{kWh} = 503\text{kWh}$$

$$\text{tier } 1 = 250\text{kWh} \times \$0.15 = \$37.50$$
$$\text{tier } 2 = 350\text{kWh} \times \$0.19 = \$66.50$$
$$\text{tier } 3 = 503\text{kWh} \times \$0.36 = \$181.08$$

Sum of tiers = $37.50 + $66.50 + $181.08 = $258.08.

43. If an object is five meters high and the solar elevation angle is 28 degrees, how long will the shadow be?

a. 5.2 meters
b. 9.4 meters
c. 6.3 meters
d. 3.8 meters

This is a trigonometry question and it is unlikely that knowing trig would be the difference in passing the NABCEP Technical Sales Exam, so do not stress if you do not know trig. However, if you have the time in your budget to spend an hour or less studying trig, you can properly answer this problem.

If you already know trig, you are ahead of most others.

We cover trigonometry in detail in the second book in this series, *Solar PV Engineering and Installation*.

Figure 8.1 Sides and angles of the trigonometry triangle.

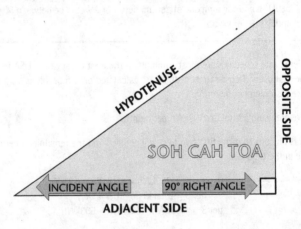

Figure 8.2 Question 43 triangle diagram.

With trigonometry, there is a relationship between the angles of a right triangle and the sides of the right triangle. All we have to do is use trig functions to extrapolate. If we know an angle and a side of a triangle, we can then figure out every other dimension of the triangle.

We label the sides of the triangle relative to the angle that we know or want to know. In the case of the solar angle, the angle of the solar elevation angle will be the angle, and the height of the object will be the side of the triangle that we know.

Figure 8.2 shows the triangle that we have for solving this problem, with the elevation angle of the sun being the angle, the **height being the side opposite** of the angle and the **distance that we want to solve for being the adjacent side** of the triangle.

The relationships of a trigonometry triangle are as follows:

Sine = opposite / hypotenuse
Cosine = adjacent / hypotenuse
Tangent = opposite / adjacent

Often people remember: SOH CAH TOA

For our exercise, since we have the angle and the opposite side of the triangle, and would like to know what the adjacent side is, we need the formula for tangent = opposite / adjacent (we are not interested in the hypotenuse).

Using our scientific calculator we find the tangent of a 28-degree angle:

Tangent of 28 degrees = 0.53

125

0.53 will be the ratio of the opposite/adjacent angles.

Since we want to solve for the adjacent side, we can do some algebra and move things around:

$$Tan\ 28 = 5m\ /\ shadow\ distance$$
$$Shadow\ distance \times Tan\ 28 = 5m$$
$$Shadow\ distance = 5m\ /\ Tan\ 28$$
$$Shadow\ distance = 5m\ /\ 0.53$$
$$Shadow\ distance = 9.4m$$

We have now solved for the distance of the shadow, which is 9.4 meters.

44. How much PV can you fit on a 44 feet wide × 32 feet north to south south-facing rooftop in portrait orientation if you are going to have a 0.5-inch space between all of the modules and if the fire department requires a three-foot perimeter on the ridge and the sides of the roof? You also need to leave a one-foot space at the bottom of the roof so that the rain will go into the gutters. The dimensions of the modules are 66 × 39 inches and they are 260 watts each.

   a. **14.3kW**
   b. 44kW
   c. 10.2kW
   d. 47kW

To determine working space of the roof in the north–south dimension we will subtract three feet on the ridge and one foot by the gutter, which means we will subtract four feet altogether.

$$32\ feet - 4\ feet = 28\ feet\ of\ working\ space$$

Since the modules are mounted portrait, we can now determine how many modules can fit on the roof by taking the long sides and remembering that there will be a half-inch between each of the modules. That means that there will be one fewer space than there are modules. One module would have no spaces, two modules would have one space, three modules would have two spaces, etc.

28 feet from the ridge to the gutter in inches is:

$$28ft \times 12in\ per\ ft = 336\ inches$$

We could come up with a mathematical equation for determining that the spaces are equal to the number of modules minus 1, but since this is a simple exercise it is easier to estimate and double check.

Here is the estimate, which will be half-inch more than required:

336"/66.5" = 5.05 modules from ridge to gutter estimate

Now let's double check:

66" modules × 5 = 330"
0.5" × 4 spaces = 2"
330" + 2" = 332"

332" is less than 336", so five modules will definitely fit from ridge to gutter in the landscape orientation.

Now we will figure out how many modules can fit lengthwise on the east to west dimension of the roof.

The roof is 44 feet wide, but we need a three-foot space on each edge of the roof, so the usable space will be:

44' – 3' – 3' = 38 feet

Just like the preceding time, we will have half-inch spaces between the modules and we will use the same method.

38' of usable space × 12" per foot = 456 inches

Here is the estimate, which will be half-inch more than required:

456" / 39.5" = 11.5 modules from edge to edge of roof

Now let's double check:

39" modules × 11 = 429"
0.5" × 10 spaces = 5"
429" + 5" =444"

444" is less than 456", so 11 modules will definitely fit from edge to edge in portrait orientation.

Now we will determine the total number of modules.

11 modules side to side × 5 modules ridge to gutter = 55 modules

And they are 260W modules, so the total power is:

$$55 \text{ modules} \times 260\text{W per module} = 14,300\text{W}$$
$$14,300\text{W}/1000\text{W per kW} = 14.3\text{kW}$$

We can fit 14.3kW of 260W modules on the roof!

Everyone works out this type of problem differently, and there are many steps. The key here is to not make an obvious mistake under pressure when you know better.

45. Which is the best type of inverter for designing a system where you can add more modules later?

    a. String inverter
    b. Central inverter
    c. Microinverter
    d. Transformerless inverter

In most situations, using microinverters makes it easier to add more modules later. There can be situations where it could be easy to add a module with a string inverter, but in general using a microinverter makes it very simple. When adding modules you do not have to worry about different orientations.

46. What would usually be a better investment?

    a. Installing thin-film PV
    b. Installing crystalline PV
    c. Switching from incandescent to CFL lighting
    d. Savings account

This book is all about PV and we would hope that PV would be the answer, but there is a saying in the industry that the most cost-effective watt that you can save with is a negawatt, which means that if you can conserve and negate using power, then you have made the best investment.

The correct answer here is switching from incandescent to CFL light bulbs. The return on investment can be in less than a year.

47. Can you mount an inverter with the top of the inverter four inches from the ceiling. If so, how will you know if you can?

    a. No
    b. Yes if it is allowed in the inverter installation manual

c. Yes if the ceiling is less than five feet

d. Yes under any circumstances

The inverter installation manual may not allow you to install an inverter within two inches of the ceiling. Often the inverter will need airflow to keep it cool. However, if the inverter installation manual will allow mounting so close to the ceiling, you may do so.

We are never allowed to install something in a way that contradicts the instructions. We also have to comply with the National Electric Code.

48. What is the minimum depth of working space that you must have facing a 5kW residential inverter mounted on a wall between the inverter and a wooden wall facing the inverter?

a. 18 inches

b. 30 inches

c. 24 inches

d. **36 inches**

The National Electric Code allows us to have a 3-foot (36-inch) working space facing the inverter when nominal voltage to ground is less than 150V or when there are no live parts, the live parts are insulated or there is an insulated wall opposite the inverter.

49. In Hawaii, if you have a low-slope rooftop with unlimited space for your PV system, what is the best tilt angle for optimal kWh/kWp/yr production?

a. **Latitude**

b. Flat

c. Latitude +15

d. Latitude −15

The old rule of thumb for PV tilt angles was latitude +15 tilt to optimize for winter, latitude −15 to optimize for summer and latitude tilt to optimize for annual production. Using software, we can better refine our tilt for optimal production and often slightly less than latitude tilt is more productive than latitude, due to weather and long summer days.

Of the given choices, latitude tilt is an obvious winner!

50. Which of the following is the most important thing you should tell your customer about having a three-degree tilt with a typical PV system?

    a. They should be prepared to have less production due to reflection

    b. They will have more wind loading than a 20-degree tilt

    c. Airflow will be better with a low tilt angle and increased voltage

    d. Expect more soiling losses

With a three-degree tilt angle, when dew and dust puddle against the frames of the modules there will be increased derating. Many solar professionals say that you need a five- to ten-degree tilt in order to avoid excess soiling losses from the buildup of soiling at the lower edge of the module frame. If you insisted on having a three-degree tilt, it would be best in a rainy climate or plan on washing the modules when there is not rain.

51. Of the following, which are the most important things to give a customer with a proposal for a PV system?

    a. String sizing calculations

    b. Size of the system and module datasheets

    c. Expected low temperature

    d. Wire sizes

String sizing calculations, expected low temperature and wire sizes are important; however, unless your customer is a solar professional, he or she will likely be overwhelmed and not understand the implications. **It is customary to give the customer the size of the system in kW and datasheets** for the modules and inverters. Another thing that you should give your customer is the expected performance of the system.

52. A utility bill has data for May and June and you are asked to give an estimate on a PV system and to size the appropriate system with the data given. Assume every month of the year has similar usage.

MAY	
1427KWH	$214
DEMAND CHARGES	$117

JUNE	
1398KWH	$210
DEMAND CHARGES	$115

If a PV system performance at the location would be 1457kWh/kWp/yr, how much of the bill could you offset and what would be the size of the system?

a. Offset $3936 with 11.6kW
b. Offset $2544 with 11.6kW
c. Offset $656 with 15.2kW
d. Offset $10,206 with 33kW

It is generally accepted that without energy storage, a PV system will offset very little demand charges.

To calculate the energy bill which we can offset with a net-metered PV system, since we only have data for two months, we will add the data from the two months and then multiply by six to extrapolate data for a year.

First, we will say that we can offset the energy usage and charges, so we will calculate the dollar amount for kWh usage for a year:

$$\$214 \text{ May} + \$210 \text{ June} = \$424$$
$$\$424 \times 6 = \$2544 \text{ per year for energy (not demand)}$$

We can estimate that the most the PV system could lower our bill by would be $2544 or slightly more, due to demand (but not much more).

To figure out how large a system would need to be to offset our entire energy bill:

$$1427\text{kWh} + 1398\text{kWh} = 2825\text{kWh}$$
$$2825\text{kWh} \times 6 = 16,950\text{kWh per year}$$

To determine how much PV it will take to make 16,950kWh, we will divide by kWh/kWp/yr:

$$16,950\text{kWh per year} / 1457\text{kWh/kWp/yr} = 11.6\text{kW}$$

We can double-check our numbers:

$$11.6\text{kW} \times 1457\text{kWh/kWp/yr} = 16,900\text{kWh per year}$$

In summary, it will take 11.6kW of PV to offset all of our energy and we will save $2544 per year.

53. Your system is making 245,000kWh per year when installed. How much power would you expect the system to make in 10 years with a conservative estimate of 1% per year degradation?

    a. 239,000kWh/yr
    b. 242,000kWh/yr
    c. 237MWh/yr
    d. 222MWh/yr

If we are going to lose 1% per year, then we will keep 99% and use the decimal form of 99%, which is 0.99 for our calculations.

To get our year 10 derating factor, we will have to multiply 0.99 by itself ten times; we can do this by calculating:

$$0.99^{10} = 0.9044$$

Next, we will multiply our first-year production by our year 10 derating factor:

$$245,000kWh \times 0.9044 = 221,578kWh$$

221,578kWh can be rounded off to 222MWh per year, which is the correct answer.

54. What is the typical process for getting an interconnection agreement for a residential 4kW system?

    a. Send in the utility paperwork and have a bidirectional meter installed before doing any work
    b. Send in utility paperwork during construction
    c. Send in utility paperwork including signed off permit after construction is completed
    d. Send in utility paperwork after utility inspection and meter change

Every utility has its own process and typically you will have to send in a signed-off permit after construction has been completed and inspected before you will get an interconnection agreement.

55. How long will it take for a typical PV system to offset its carbon footprint, including the PV inverter and balance of systems?

   a.   Less than 6 months
   b.   **1 to 2 years**
   c.   5 years
   d.   10 years

The typical PV system's carbon footprint will be offset in about 1–2 years of operation. This includes the process for refining silicon, making cells, modules, inverters, etc.

56. What will you do to determine if you can install a PV system on an industrial metal building in Buffalo, New York?

   a.   Verify that the racking system can support the system on the building by reading the installation manual
   b.   **Hire a structural engineer**
   c.   Get permission from the Buffalo Chamber of Commerce
   d.   Get an interconnection agreement

Especially in places with snow, it is important to see if the building can handle the extra weight of a PV system. You do not want to spend a lot of time designing and selling a PV system if in the long run you cannot install it on the building because of structural concerns that would prohibit installing the system due to high costs of reinforcing the building or not being able to install the system at all. You need to know what you are working with. Hiring a structural engineer from the beginning can save a lot of money in the long run.

57. Your customer is an ecologically conscious consumer and you can tell that she is interested in doing the right thing. What should you emphasize when generating a proposal for her?

   a.   **Carbon offsets**
   b.   Nitrogen fixing
   c.   How the PPA can sell tax credits on Wall Street
   d.   How you calculate the string length

An ecologically conscious consumer will be interested in how a PV system will offset greenhouse gasses, how many trees it will save and how many tons of coal it will leave in the ground. Many proposal software programs will have an option to include this material for customers that are interested in more than saving money. The ecologically conscious consumer is often the first to go solar!

58. You are installing a system that should make 1360kWh for every kW installed. The system has had some unforeseen losses since the modules are in a place where it is very dusty and the customer is not cleaning the modules, which leads to 12% of extra losses. If the system makes 10,100kWh per year after the losses from soiling, how large is the PV system?

    a. 4.8kW
    b. 8.4kW
    c. 9.5kW
    d. 10.6kW

If we should usually get production of 1360kWh/kWp/yr and we are going to lose 12% more, then we need to derate 1360 by 12%.

To turn 12% into a derating factor, we want to determine how much we keep.

$$100\% - 12\% = 88\%$$

We are going to move the decimal two places to the left in order to turn the percentage into a decimal derating factor of 0.88.

Now to derate the 1360kWh/kWp/yr:

$$1360\text{kWh/kWp/yr} \times 0.88 = 1197\text{kWh/kWp/yr}$$

If a system makes 10,100kWh per year while making 1197kWh/kWp/yr, to determine the kW rating of the system we divide the kWh/yr by the kWh/kWp/yr:

$$10,100\text{kWh per year} / 1197\text{kWh/kWp/yr} = 8.44\text{kW}$$

To double check, we can multiply:

$$8.44\text{kW} \times 1197\text{kWh/kWp/yr} = 10,100\text{kWh per year (rounding off)}$$

Therefore, we can confirm that our system size is 8.44kW.

59. With southeast- and south-facing arrays of thirteen 240W modules each, what is the best inverter combination for maximum production? All inverters listed have a single MPP.

    a. Two 3kW inverters
    b. One 7kW inverter

c. Three 2.5kW inverters

d. One 2kW inverter and one 8kW inverter

Each array will have 13 modules and it is reasonable to assume that they could all fit on a single PV source circuit. If the inverter was a 600V inverter, then:

$$600V / 13 \text{ modules} = 46V$$

For a 240W module to get to 46V, it would have to be very cold!

Since each array has 13 modules and the modules are 240W, the power of each array would be:

$$240W \times 13 \text{ modules} = 3120W$$

With 3kW inverters, the PV to inverter ratio is:

$$3120W / 3000W = 1.04$$

This means that the ratio is low enough that the inverter will never clip or have the ability to make more power than the inverter can put out. That is good for using 3kW inverters.

With one 7kW inverter that has a single MPP, it would not be bad, but the MPP voltage of each array would be slightly different due to different orientations causing different temperatures on each array.

With three 2.5kW inverters, there would have to be a PV source circuit that had multiple orientations since three inverters would have to be split into two orientations with equal modules in each array.

A 2kW and an 8kW inverter would not be optimal, because the 2kW inverter would not be enough for one of the directions. If I had a 2kW and an 8kW inverter for this job, I would use the 8kW inverter and take the 2kW inverter home for another job.

The best option is two 3kW inverters, one for each array. Each array would be tracked individually and the PV is slightly more than the inverter power, which is just right.

If we had a single 6kW inverter with dual MPP inputs, then I would chose the 6kW inverter, since it would be easier to install the ac wiring on a single inverter and the separate MPPs would act the same as having two separate inverters.

60. With a 225A busbar and a 175A main breaker, what is the largest inverter you can install using the 120% rule on a 240V service?

    a.   22kW
    b.   14kW
    c.   5,200W
    d.   **18.2kW**

Calculate 120% of busbar:

$$225A \times 1.2 = 270A$$

Subtract main breaker:

$$270A - 175A = 95A$$

Divide by 1.25 correction factor:

$$95A / 1.25 = 76A$$

Instead of dividing by 1.25, we can multiply by 0.8 for same answer:

$$95A \times 0.8 = 76A$$

Calculate power (amps × volts = power):

$$76A \times 240V = 18,240W \text{ inverter}$$

Change W to kW:

$$18,240W / 1000 = 18.2kW \text{ inverter}$$

61. What is the least expensive way to install a 4kW inverter on a house with a 100A busbar and a 100A main breaker?

    a.   120% rule load-side connection
    b.   **Supply-side connection**
    c.   Upgrade service panel to 200A
    d.   Line side tap

A load-side connection would be most cost-effective. With this inverter, we are barely overpowered to be able to do a load-side connection using the 120% rule. We can calculate this here:

4kW inverter current at 240V is:

$$4000W / 240V = 16.67A$$

According to the 120% rule, we will use 125% of the inverter current in our calculations. 125% of our current is:

$$16.67A \times 1.25 = 20.8A$$

Then we will do the 120% rule calculations with the busbar and the main breaker:

$$100A \text{ busbar} \times 1.2 = 120A$$
$$120A - 100A \text{ main breaker} = 20A \text{ inverter allowance}$$
$$20.8A > 20A$$

(According to the 2011 NEC and earlier, we would have had to use 25A in the calculations instead of 20.8A, because we had to round up to the next common overcurrent protection device size instead of using 125% of inverter current in our calculations.)

Since 125% of the inverter current is greater than 20A, then we are not allowed to do a load side 120% rule connection.

The next most cost-effective way of connecting solar would be using a supply-side connection, which has been incorrectly called a line side tap. It is not a tap, since it cannot follow the tap rules, since there is no overcurrent protection on the supply side of the connection.

For a supply-side connection, we are connecting between the meter and the first overcurrent protection device (main breaker). With this type of connection, we are limited by the size of the service, which is always more than we would ever want to connect. Since the main breaker is 100A, we would have to have service entrance conductors that are at least 100A and they could handle 125% of the inverter current.

$$(100A / 1.25) \times 240V = 19.2kW$$

We could have an inverter that could be at least 19.2kW and perhaps more if the service entrance conductors were larger. There is typically no good reason why we would want a 19.2kW inverter on a 100A service. That would make more energy than a 100A service would use.

It would cost more money to do a panel upgrade than a supply-side connection, so the correct answer is to do a supply-side connection.

Many solar companies will sell the panel upgrade and install a 200A panel-board for a price, which can be a good idea on an old house, just not the most cost-effective solution.

62. Which will have better energy production?

    a. Ground mount
    b. Pole mount
    c. Flush roof mount
    d. Ballasted roof mount

If all other factors, such as tilt, azimuth, shading and other deratings, are the same, the only difference we would have here is airflow. A pole-mounted system will have the most airflow, since it is exposed the most. Next would come a ground-mounted and then the roof-mounted systems. The systems that typically have the least amount of airflow are building integrated PV (BIPV) systems. The increased airflow on the pole-mounted systems will lead to cooler operating temperatures, more voltage, power and energy.

63. You are selling a ground mount at a school and the school is concerned about children getting electrocuted by the array. What should you tell the school?

    a. There will be a fence around the array
    b. Children would get shocked, not electrocuted in a worst-case scenario
    c. The connectors between the PV modules are the locking and latching type and will not allow the children to get shocked
    d. There is more danger from changing a light bulb than from a PV array

The difference between electrocution and shock is that electrocution is death by shock. Either death or shock is possible in a PV array. Also, with higher dc voltages in the typical solar array, it can be more dangerous to get shocked by a PV array than by changing a typical light bulb.

A ground mount should have the wiring behind it made inaccessible. This is most often done with a ground mount having a fence around it.

64. In a location at 45 degrees latitude, you have a roof that is sloped 30 degrees to the south. If you are going to mount PV on that roof, what would be the best way to mount the PV?

a. Reverse tilt on north side of roof, so that the PV is tilted 45 degrees to the south

b. Extra tilt with PV mounted on the south side of roof, so that PV is tilted at 45 degrees to the south

c. PV tilted on the south side of the roof to 40 degrees south

d. **PV installed with the slope of the roof**

In this situation, long ago when PV was very expensive, some people would have tilted the PV modules to latitude tilt on a sloped roof. Now that PV is less expensive, almost everyone would install the PV flush mounted with the roof tilt and azimuth. The production at 45 degrees latitude with a 45- or a 40-degree tilt will be very similar to production at a 30-degree tilt.

65. Who gets the final say in whether you can build on a rooftop that is questionably stable when the weight of a PV system is added?

a. **AHJ**

b. Racking manufacturer

c. Structural engineer

d. EPC contractor

The authority having jurisdiction (AHJ) will always have the final say with any project. When we refer to the AHJ, we typically mean someone working for the building department, but it can mean the utility, federal government, state government or other authority.

The AHJ will often require a "wet stamp" from a structural engineer, but they do not have to require a structural engineer to review the project, since it is up to the AHJ.

The AHJ is the final authority and they can require documentation from the racking manufacturer, a structural engineer and from the EPC contractor, but it is up to the AHJ to decide what the requirements are.

EPC is **e**ngineering, **p**rocurement and **c**onstruction and the EPC contractor is typically who runs the job, though every job can be run differently.

66. What would happen if the installation crew used the smallest wire required by the code from an inverter to the point of common coupling, which was 2000 feet away?

a. **The inverter would likely anti-island**

b. The inverter would stay on, but make less

    c.   The wire would burn

    d.   The overcurrent protection devices would not function

As long as the wire was up to code, it should not be a danger; however, the big problem is going to be severe voltage drop. When there is voltage drop on an inverter output circuit, since the power is coming from the inverter to the point of connection, then the voltage at the inverter has to be higher in order for the power to go in the direction of the connection. With severe voltage drop from using the smallest acceptable wire and going a large distance, the inverter would undoubtedly go out of voltage range and anti-island. In the industry, we call this effect voltage rise, since the voltage at the inverter will go up, as there is more current. With a 2000-foot distance, it would take at least 4000 feet of wire to get there, since every circuit is a circle (there and back).

Voltage drop in a PV system is not a safety issue, but it does lead to poor performance.

67. With 15% efficient PV, what would be the most PV that you could fit on a rooftop that is 390 square feet?

    a.   3.1kW

    b.   4.2kW

    c.   **5.4kW**

    d.   6.3kW

15% efficient PV means we will be producing 15% of STC, which is 1000W per square meter.

$$15\% \text{ of } 1000W/\text{square meter} = 150W \text{ per square meter}$$

Since we need to figure out how much PV will fit on an area that is measured in square feet, we need to convert square feet to square meters.

Most people round off and call 1 meter 3 feet; however, we could also round it off to two significant digits to 3.3 feet or three significant digits to 3.28 feet.

Another way is if we know that there are 2.54cm to an inch.

$$2.54cm \times 12 \text{ inches} = 30.48cm \text{ to a foot}$$

Since a meter is 100cm, we can move the decimal two places:

$$30.48cm = 0.3048m = 1 \text{ foot}$$
$$1 \text{ foot} / 0.3048m = 3.28ft/m$$

There are a lot of ways to go about it and it is good to practice playing with numbers.

1 square meter = 3.28 feet × 3.28 feet = 10.76 square feet per square meter

If you guessed at 11 square feet per square meter, you would be accurate to 2 significant digits, which is not bad and might get you the right answer.

Memorize 2.54cm per inch and you can't go wrong, because that is accurate!

Since we have 150W per square meter, then we have:

150W/10.76 square feet = 13.9 watts per square foot

Since we have 390 square feet:

390 square feet × 13.9W per square foot = 5421W

5421W is about 5.4kW.

Some people are conservative and estimate 10W per square foot, since we will not be able to fill up every space on the roof and it is very quick to calculate 10W per square foot.

68. What is the main reason to know the low temperature for designing a PV system?

    a. Low temperatures can freeze solar cells
    b. Low temperatures cause voltage to decrease
    c. Low temperatures cause current to increase
    d. **Low temperatures cause voltage to increase**

We have to know what the low temperature is, because in colder places voltages will get higher. When the voltage gets higher, we will not be able to put as many modules in series without causing damage to the inverter or voiding the inverter's warranty.

69. You sell a 100kW PV system and the customer complains that the system is not performing close to 100kW. You go to see him and the system is operating on a summer day and the cell temperature is 65C. The irradiance in the plane of the array is 900W per square meter and other derating factors are determined to be 0.77. What would you expect the output of the inverter to be if the temperature coefficient for power is –0.5%/C?

    a.   **55kW**
    b.   70kW
    c.   80kW
    d.   33kW

Besides the derating factor of 0.77, there will be a derating factor for temperature and another derating factor for irradiance.

To calculate the derating factor for irradiance, we will compare 900W per square meter to STC conditions of 1000W per square meter:

$$900/1000 = 0.9 \text{ derating for irradiance}$$

To calculate the derating for temperature, we will find the difference in temperature from STC, which is 25C:

$$65C - 25C = 40C \text{ difference}$$

Then we will multiply the temperature coefficient of power by the difference in temperature from STC:

$$40C \times -0.5\%/C = -20\% \text{ power}$$

If we lose 20%, we keep 80% and our derating factor will be 0.8.

Now we have three derating factors and we can multiply them together to get an inclusive derating factor:

$$0.77 \times 0.9 \times 0.8 = 0.55 \text{ derating factor}$$

To calculate what the 100kW array should be making on that warm day:

$$100kW \times 0.55 = 55kW$$

We would expect the array to provide about 55kW coming out of the inverter.

70. With a 3kW inverter that can have five to ten 175W modules in series, you would like to install 18 modules all facing the same direction. Five of the modules will be shaded for an average of two hours per day. The inverter has a single MPP input. What would be the best configuration of the following?

    a.   Two PV source circuits of nine modules in series
    b.   One PV source circuit of eight and another of ten

c. One source circuit of five, another of six and another of seven

d. Three PV source circuits of six

Two source circuits of nine would be the best if there were no shading; however, when there is shading of the five modules, we would have about 50% production for two hours per day, which would be best avoided.

With source circuits of different lengths on a single MPP inverter, we would not be operating on the MPP of both PV source circuits, and that is not recommended in any case, with or without shading. Visualize two or three different IV curves that all have maximum power voltage at different points, which is not good.

With three source circuits of six each, if we keep all of the shaded modules on the same string, we would have two-thirds of the array working efficiently at all times and one third of the array operating like it was in the shade for two hours per day. This is the best solution.

30 bonus practice questions

1. Of the following, which is the largest inverter that you can interconnect with a load-side connection on a busbar that is 200A, with a main breaker that is 200A? There is also a 100A subpanel breaker and a 40A load breaker connected to the main service panel. You want to connect the solar to the main service panel. Assume that the voltage of the inverter is 240V single-phase.

   a. 9kW
   b. 7.5kW
   c. 11kW
   d. 5kW

2. If you were planning on connecting solar to a feeder between a 200A panelboard, with an existing 100A feeder breaker, and a 120A rated conductor connecting the feeder breaker to the 125A rated subpanel, and you wanted to connect a 5kW inverter to the feeder, in order to avoid replacing the feeder, what would be the best solution, or is there a solution? Assume that the voltage is 240A.

   a. Putting an OCPD on the feeder between the feeder OCPD and the inverter/feeder connection point.
   b. Putting an OCPD on the feeder between the inverter/feeder connection point and the 125A subpanel.
   c. There is no solution; the feeder must be replaced with a larger feeder to accommodate the inverter currents.
   d. Nothing has to be modified and the solar can be installed with no modifications.

3. If utility rates are expected to go up 3% per year, how much would the rates go up over a 20-year period?

   a. 60%
   b. 160%
   c. 81%
   d. 88%

4. With an ungrounded 6kW inverter that has two MPP inputs, if there are 14 modules on each string, what would happen to production if one of the modules is shaded in the middle of the day? About how much production would you expect to lose from the shading?

   a. 50%
   b. 25%
   c. 8%
   d. 4%

5. At 50 degrees north latitude, of the following, which would be the way to get the most kWh out of a flat roof in one year's time? Assume you are using thin-film PV laminates.

   a. 50 degrees tilt angle
   b. 30 degrees tilt angle
   c. 0 degrees tilt angle
   d. 35 degrees tilt angle

6. A utility has a tiered utility rate schedule. For the first 150kWh per month customers pay 10 cents per kWh and for the next 200kWh the price is 15 cents per kWh. The top and third tier is priced at 20 cents per kWh. If the customer uses 550kWh per month, what would be the system size for the best return on investment (ROI). Assume that the PV system is priced at $2.25 per watt installed, no matter how large the system is, and that the system will make 1339kWh/kWp/yr.

   a. 1.8kW
   b. 2.9kW
   c. 3.6kW
   d. 4.9kW

7. If energy prices are 5 cents per kWh from 6 a.m. to 6 p.m., 30 cents per kWh from 6 p.m. to midnight and 22 cents per kWh from midnight to 6 a.m., what would be the best investment opportunities for the customer to look into, assuming it costs 10 cents to make energy with PV and 10 cents per kWh to shift loads with energy storage?

   a. PV system
   b. PV system with energy storage
   c. Energy storage system
   d. Diesel generator with PV

8. If you have PV with an open circuit voltage of 38V and an inverter with a maximum input voltage of 1000V in a location that will have a low design temperature of –32C, assuming a typical temperature coefficient of Voc of –0.32%/C, what would be the maximum number of modules in series?

   a. 18
   b. 22
   c. 24
   d. 20

9. There are 2.1 pounds of CO_2 put into the air per kWh of energy produced by coal. You are calculating how much CO_2 you would offset over a 25-year lifespan by installing each solar module. The Imp of the solar module is 8A and the Vmp is 30V. If you install the PV in a location where it will average 1333kWh/kWp/yr during the next 25 years, then how many pounds of CO_2 will you prevent from going into the air for each solar module installed?

   a. 672 pounds
   b. 16,800 pounds
   c. 10,000 pounds
   d. 33,393 pounds

10. A 100MW solar project is expected to make 90% of its year 1 production on year 20. The first year's production came out to be 1492MWh/MW/yr. What is the expected year 20 income if the FIT rate is 10 cents per kWh?

   a. $13,428,000
   b. $14,920,000
   c. $149,200,000
   d. $134,280,000

11. A main service panel is rated for 125A and the main breaker is rated for 100A. If it is a 240V ac service, what is the largest inverter of the following that may be connected with a supply-side connection?

    a. 5kW
    b. 10kW
    c. 15kW
    d. 17kW

12. A 255W monocrystalline module is used for a 2MW system. What would be the expected performance of the array in 30 years if the degradation rate were 0.5% per year? The system on year 1 made 1452kWh/kWp/yr.

    a. 2490MWh/MWp/yr
    b. 25,0000kWh/yr
    c. 2500MWh/yr
    d. 7600MWh/yr

13. Which system will perform better, assuming that there is no shading and the orientation of each system is the same?

    a. Flush roof mount
    b. Ballasted roof mount without wind deflectors and extra ballast
    c. Ballasted roof mount with wind deflectors
    d. Ground mount

14. A PV module measures 1678mm × 998mm and will produce 255W at STC. Assuming full coverage of the roof area, about how much PV could fit on 23m² of rooftop?

    a. 3.2kW
    b. 35,000W
    c. 28kW
    d. 2.8kW

15. Who interprets the National Electric Code?

    a. Contractor
    b. State licensing board
    c. Federal licensing board
    d. AHJ

16. What is the least expensive way to properly connect a 4kW PV array to a house with a 100A panelboard and a 100A main breaker? Assume that there are 100A of loads on the busbar, but there are still extra spaces available.

    a. 4kW inverter load-side connection
    b. 3.8kW inverter load-side connection
    c. 4kW inverter supply-side connection
    d. 4kW inverter line-side tap

17. An industrial building has a utility bill for $100,000 per year for energy, $1,200 for meter charges and $33,000 per year for demand charges. You are planning to install a net-metered PV system without energy storage. If the customer uses 701,000kWh per year, which is the most that you could reduce the customer's bill?

    a. $100,000
    b. $133,000
    c. $134,200
    d. $89,000

18. There are many PV systems that are mounted close to flat (zero-degree tilt angle). If you were going to mount a system flat, what would be an important distinction to make when purchasing equipment for the flat-mounted system that would give you better performance, that would not give significantly better performance with a tilted array?

    a. Ballasted system
    b. PV laminates
    c. Framed modules
    d. Ungrounded inverter

19. You are using a subpanel to combine inverter output circuits. The subpanel is rated for 125A. What are the maximum inverter currents that could combine on the subpanel? Also consider that there will be a single 15A load breaker on the subpanel used for a monitoring device.

    a. 40A
    b. 88A
    c. 25A
    d. 45A

20. For a stand-alone system that will be used year round, what would be the best tilt angle at 40 degrees latitude, in a location that has no significant shading? You would like the system to have significant production year-round.

    a.   40-degree tilt
    b.   25-degree tilt
    c.   30-degree tilt
    d.   55-degree tilt

21. If you are installing a small PV system on a roof with many different angles, which would be the best solution? The customer also plans on buying an electric car in two years and expanding the PV system.

    a.   Central inverter
    b.   String inverter
    c.   Power optimizer/inverter combination
    d.   Battery inverter

22. What is closest to the slope of a 4:12 roof?

    a.   15 degrees
    b.   18 degrees
    c.   20 degrees
    d.   22 degrees

23. According to the 120% rule, what is the largest inverter of the following that can be installed on a 120/240V building with a 225A busbar and a 175A main breaker?

    a.   15kW
    b.   7kW
    c.   8kW
    d.   18kW

24. An investment tax credit is better financially for the system owner if compared to another system composed of the same equipment if. . .?

    a.   The system produces 50% more energy every year
    b.   The system costs 50% more
    c.   The system does not break on year 1
    d.   The system is installed in a cooler location

25. What happens with a typical dc-coupled grid-tied battery backup PV system when the grid is working and the batteries are charged?

    a. The charge controller will feed the grid
    b. A backup circuit will switch on and feed the loads with the PV system
    c. The batteries will feed the grid
    d. The diversion load will be the grid

26. You are measuring azimuth with a magnetic compass in California and the magnetic declination is 14 degrees; what would be the true azimuth of the white, magnetic south needle of the compass?

    a. 180 degrees
    b. 0 degrees
    c. 194 degrees
    d. 166 degrees

27. You are looking to install an ungrounded string inverter with two separate MPPs. The inverter will take between 5 and 14 modules in series for the particular temperature conditions and the specified monocrystalline 265W modules. You would like to put two strings of ten modules in series on the roof, but your sales team is worried about a chimney that will shade one PV module for about two hours per day. When that module is totally shaded, about how much total production will be lost?

    a. 5%
    b. 50%
    c. 10%
    d. 25%

28. If you were going to check the service voltage on a large industrial building in Nevada, what would you expect the voltage to be?

    a. 120/208V
    b. 120/240V
    c. 277/480V
    d. 1000V

29. What size overcurrent protection device should you have at the output of a 4kW inverter connected to the supply side of the main breaker on a house?

   a. 20A
   b. 25A
   c. 30A
   d. 40A

30. If a 5kW PV system were to last for 50 years and produce an average of 1400kWh/kWp/yr, how much money would the PV system save if the average adjusted for inflation price of a kWh over 50 years is 30 cents per kWh?

   a. $50,000
   b. $350,000
   c. $77,000
   d. $105,000

30 bonus practice questions with detailed explanations

Chapter 10

1. Of the following, which is the largest inverter that you can interconnect with a load-side connection on a busbar that is 200A, with a main breaker that is 200A? There is also a 100A subpanel breaker and a 40A load breaker connected to the main service panel. You want to connect the solar to the main service panel. Assume that the voltage of the inverter is 240V single-phase.

   a. 9kW
   b. 7.5kW
   c. 11kW
   d. 5kW

This is a trick question. At first you might want to apply the 120% rule, but we can add more according to the 2014 NEC sum rule.

According to the sum rule, we have to put a sign at the panelboard of the interconnection that explains the sum rule perfectly:

> **WARNING:**
>
> THIS EQUIPMENT FED BY MULTIPLE SOURCES.
> TOTAL RATING OF ALL OVERCURRENT DEVICES,
> EXCLUDING MAIN SUPPLY OVERCURRENT DEVICE,
> SHALL NOT EXCEED AMPACITY OF BUSBAR.

Therefore, if the busbar is rated at 200A and the breakers on the busbar are a 100A subpanel breaker and a 40A load breaker, what we can do is subtract the breakers on the busbar from the busbar rating. Note here that for this rule we ignore the main breaker, since the busbar will be protected by the breakers on the load side of the main breaker.

Breakers on busbar (excluding the main breaker):

100A subpanel + 40A load breaker = 140A

Now we subtract the 140A from the rating of the busbar to get our solar breaker size.

200A busbar − 140A = 60A solar breaker

Since the breaker is sized to be a maximum of 125% of the inverter current, then we can either divide the 60A by 1.25 or multiply the 60A by 0.8, which is the same thing.

60A solar breaker × 0.8 = 48A solar inverter

Since the inverter is 240V on the ac side then we will simply multiply volts by amps to get our watts:

240V × 48A = 11,520W = 11.5kW

The question asked us to pick the largest of the following, so **the correct answer is C, 11kW**.

If you applied the 120% rule (if you were applying the 2011 NEC or earlier), then you would have done the following:

200A busbar × 1.2 = 240A allowance
240A allowance − 200A main breaker = 40A solar breaker
(or 125% of inverter current)
40A × 0.8 = 32A inverter current max
32A inverter current × 240V = 7689W = 7.7kW

In this case you would have selected B, as the 7.5kW inverter would have been the largest of the choices.

If you were trying to avoid a supply-side connection so you did not have to deal with pulling the meter or waiting for the utility to pull the meter – since it can get very complicated with certain utilities – you could place many of your loads that were on a main panel on a subpanel and then make room to apply the sum rule, avoiding a supply-side connection.

Supply-side connections require turning off the power. For larger buildings, where the power does not go through the meter and the meter works with CTs (current transducers, which are like clamp-on amp meters), the utility will have to shut off the power.

For houses, the meter is typically pulled. In utility districts where solar is prevalent, there is a tendency for the utility to let the solar contractor pull the meter in order to save time. In utility districts where the utility is not as experienced with solar, there is a tendency for the utility to require an appointment for the utility to come out to pull the meter, which adds considerable costs due to the time, scheduling and waiting for the utility worker to show up to pull the meter. Utilities often suspect that someone it trying to steal power when the meter is removed.

2. If you were planning on connecting solar to a feeder between a 200A panelboard, with an existing 100A feeder breaker, and a 120A rated conductor connecting the feeder breaker to the 125A rated subpanel, and you wanted to connect a 5kW inverter to the feeder, in order to avoid replacing the feeder, what would be the best solution, or is there a solution? Assume that the voltage is 240A.

    a. Putting an OCPD on the feeder between the feeder OCPD and the inverter/feeder connection point.

    b. **Putting an OCPD on the feeder between the inverter/feeder connection point and the 125A subpanel.**

    c. There is no solution; the feeder must be replaced with a larger feeder to accommodate the inverter currents.

    d. Nothing has to be modified and the solar can be installed with no modifications.

The answer is B, the feeder can be protected by placing an OCPD on the feeder between the inverter/feeder connection point and the 125A subpanel. This way the feeder is protected from the currents of the inverter and from the utility coming from the 100A subpanel breaker.

If the inverter were very small or if the conductor going to the subpanel and the subpanel were larger, then we would not need to modify anything.

We would have to be protected from the 100A subpanel breaker plus 125% of the inverter current.

To calculate 125% of the inverter current, we will first calculate what the inverter current is by dividing power by voltage:

$$5000W \text{ inverter} / 240V \text{ ac} = 20.8A$$

Then to get 125% of the inverter current:

$$20.8A \times 1.25 = 26A$$

In this case, we would need a conductor and a subpanel busbar that would be able to handle the 100A subpanel breaker plus the 26A allowance from the inverter.

$$100A \text{ subpanel breaker} + 26A \text{ inverter allowance} = 126A \text{ required}$$

Since the conductor is rated for 120A and the subpanel busbar is rated for 125A, neither can handle the currents without upsizing both the subpanel and the feeder from the inverter connection to the subpanel.

The feeder between the 100A feeder breaker and the inverter connection point is fine the way it is and there would never be a reason to oversize this part of the feeder with this example. Only the portion of the feeder which is subject to the currents from the 100A subpanel breaker and the inverter combined going to the subpanel needs extra protection.

3. If utility rates are expected to go up 3% per year, how much would the rates go up over a 20-year period?

   a. 60%
   b. 160%
   c. **81%**
   d. 88%

To determine how much the rates will go up, we have to multiply the decimal equivalent of a 3% increase by itself 20 times. The way to do this on a calculator is by first converting 3% to a decimal:

$$3\% / 100\% = 0.03$$

We then add 1 to this number, making it a 3% increase:

$$1 + 0.03 = 1.03$$

Instead of pressing too many buttons and multiplying 1.03 by 1.03 twenty times, we can use the X^y button.

- If 1.03 is on your calculator, then press X^y
- Next press 20, for 20 years.

The number we get will be 1.81, which represents an 81% increase.

If you multiply something by 1.81 you will get an 81% increase, so **the correct answer is C, 81%.**

4. With an ungrounded 6kW inverter that has two MPP inputs, if there are 14 modules on each string, what would happen to production if one of the modules is shaded in the middle of the day? About how much production would you expect to lose from the shading?

    a. 50%
    b. 25%
    c. 8%
    d. **4%**

There are many people out there who think that if you shade a single module, that it will take out the entire PV source circuit (string). In reality, if there were a single shaded module on a string inverter, the bypass diodes will bypass the shaded module and the string will act exactly as if there were a single module fewer.

With an inverter with two separate MPPs and two separate strings, when a single module out of 28 total is shaded, we will only lose 1/28th of the production.

$$1 \text{ module} / 28 \text{ modules} = 0.04$$
$$0.04 \times 100\% = 4\% \text{ loss in production}$$

The answer is D, only a 4% loss in production.

If the arrays were on an inverter with a single MPP, the losses would be slightly more, since the array would be unable to MPP strings of different lengths at once.

If module-level electronics were used here, such as microinverters or power optimizers, the shaded module would not be bypassed and would work as well as a module in the shade can work. In the shade, we may expect 5–20% performance on the shaded module, depending on the characteristics of the shade. 10% performance on a shaded 250W module would be 25W.

5. At 50 degrees north latitude, of the following, which would be the way to get the most kWh out of a flat roof in one year's time? Assume you are using thin-film PV laminates.

    a. 50-degree tilt angle
    b. 30-degree tilt angle
    c. **0-degree tilt angle**
    d. 35-degree tilt angle

The best way to get maximum kWh out of a roof, if we ignored the price of the system, is to mount the PV flat. With a flat system, we could cover the entire roof without having to leave spaces for inter-row shading, and catch as many photons as possible. **The correct answer is C, 0-degree tilt angle.**

The reason we do not always cover rooftops with flat-mounted PV is that flat-mounted PV does not produce as much energy per kW. Flat-mounted PV will not produce as many kWh/kWp/yr as PV tilted at latitude, 30 degrees or anything in between. Flat-mounted PV will, however, have a better ground cover ratio (GCR), especially at higher latitudes where more space is required between tilted rows of modules.

A problem with flat-mounted PV is soiling on PV with frames. PV laminates do not have frames, so the soiling problems will be much less than with framed modules. Another benefit for thin-film modules is that they work better under low light levels. On cloudy days, flat-mounted PV will do better than tilted PV.

Another benefit for a lower tilt angle or a flat-mounted array is that it will have less trouble with wind loading. The more tilted a PV module is, the more it will take to keep it from blowing away. Ballasted PV systems require less ballast weight when they are sloped less.

Also, in snow country, tilted PV will keep the snow from blowing away; this is called snowdrift. With a lower slope or a flat-mounted array, snowdrift becomes less of a problem. With heavy snowdrift building up behind tilted modules, the snow weight could collapse a rooftop.

After all this is said, it is still better to tilt PV in most cases, but you can see the benefits to having a lower tilt angle. Often at 50 latitude, low slope roof racks are tilted 10–20 degrees.

6. A utility has a tiered utility rate schedule. For the first 150kWh per month customers pay 10 cents per kWh and for the next 200kWh the price is 15 cents per kWh. The top and third tier is priced at 20 cents per kWh. If the customer uses 550kWh per month, what would be the system size for the best return on investment (ROI). Assume that the PV system is priced at $2.25 per watt installed, no matter how large the system is, and that the system will make 1339kWh/kWp/yr.

    a. **1.8kW**

    b. 2.9kW

    c. 3.6kW

    d. 4.9kW

The best return on investment would be to only offset the most expensive kWh. If the customer uses 550kWh per month total, first we will have to subtract the kWh that are not in the most expensive tier.

- 10 cent tier = 150kWh per month
- 15 cent tier = 200kWh per month

Total of lower tiers:

$$150kWh + 200kWh = 350kWh \text{ per month}$$

To see how much is in the most expensive 20 cent tier, we will subtract the lower two tiers from the total energy usage in a month.

$$550kWh \text{ total} - 350kWh \text{ lower tiers} = 200kWh \text{ per month in high tier}$$

To size our system, we will multiply 12 months by the monthly usage:

$$12 \text{ months} \times 200kWh \text{ per month high tier} = 2400kWh \text{ per year}$$

If our production in this location is going to produce 1339kWh/kWp/yr, then we divide the amount of energy we want to make by the kWh/kWp/yr:

$$2400kWh / 1339kWh/kWp/yr = 1.8kW \text{ of PV}$$

The answer is A, 1.8kW of PV is needed to make the best return on investment.

In reality, installing such a small system would have a greater cost than a larger system due to getting the trucks rolling, filing for a permit, interconnection fees and labor, etc. Most solar companies would not spend their time installing such a small system.

Also, the ROI for offsetting the other tiers may be a good investment and solar companies may chose to offset the different tiers.

This example had energy usage assumed to be the same throughout the year, which is not the case in most places.

7. If energy prices are 5 cents per kWh from 6 a.m. to 6 p.m., 30 cents per kWh from 6 p.m. to midnight and 22 cents per kWh from midnight to 6 a.m., what would be the best investment opportunities for the customer to look into, assuming it costs 10 cents to make energy with PV and 10 cents per kWh to shift loads with energy storage?

    a. PV system
    b. PV system with energy storage
    c. **Energy storage system**
    d. Diesel generator with PV

Since the high rates are in the evening and night time and the low rates are during the day time, and the lower rates are less expensive than making energy with PV, this job does not have a financial reason to have PV from the information given, unless the economics change during the life of the PV system.

Low energy prices are the enemy of PV salespeople. There are many places where fossil fuel energy is subsidized more than PV energy, and in those places there is often not an economic reason to get PV in the short term.

If an energy storage system can take the 5 cent per kWh energy to charge the batteries and use that energy when prices are 30 cents per kWh, that would be a difference of 25 cents per kWh. Since the battery costs are calculated to be 10 cents per kWh, then we can make 15 cents per kWh. This question did not say if the round-trip efficiency for the batteries was included in the 10 cents of the battery costs. Even if it were not included in the costs, it would still be the only way to make money given the information above. **Therefore, the correct answer is C, an energy storage system is the only way to make money in this situation**.

We should all know that the most expensive of the options above would be using diesel. It would not make sense to use PV or diesel with PV when energy is so cheap when the sun is up.

These are all financial reasons and we left out the social reasons for getting PV. It is good for the environment, the neighbors like it, it decreases CO_2 released into the atmosphere and it will make people feel good for doing the right thing. Other reasons for decreasing our dependence on fossil fuels are that the polluting effects cause a rise in healthcare costs.

8. If you have PV with an open circuit voltage of 38V and an inverter with a maximum input voltage of 1000V in a location that will have a low design temperature of –32C, assuming a typical temperature coefficient of Voc of –0.32%/C, what would be the maximum number of modules in series?

   a. 18
   b. 22
   c. 24
   d. 20

If we can figure out the Voc of the 38V module when it gets really cold, we can then see how many we can put in series without going over 1000V.

To figure out the cold temperature Voc, we first need to see how much colder it is than when the modules are rated and tested at 25C. So the difference between 25C and –32C is going to be just like adding 25 and 32 since they are each on opposite sides of zero.

$$25C + 32C = 57C$$

If we look at the units of the temperature coefficient for Voc, they are %/C, which means if we multiply the coefficient by the temperature change we will get a percent increase in voltage.

$$57C \times -0.32\%/C = 18.24\% \text{ increase in Voc}$$

As you may notice here, we are using a slightly different method than in the book, so that the concept is imprinted into your mind. This way, we are reasoning out exactly what is happening and are ignoring double negatives with the degrees C and the coefficient, and are using common sense that tells us that the voltage will go up when the temperature goes down.

Since we have an 18.24% increase in voltage, that means we will have 18.24% more than 38Voc. We could calculate what 18.24% of 38V is and then add it to 38V.

First, we turn 18.24% into a decimal:

$$18.24\% / 100\% = 0.1824$$

Then we multiply by 38V to get 18.24% of 38V:

$$38V \times 0.1824 = 6.9V \text{ increase in voltage}$$

Then we add the increase in voltage to the voltage:

$$38V + 6.9V = 44.9Voc \text{ when cold}$$

Now to figure out how many modules go in series to stay under 1000V by dividing 44.9V into 1000V:

$$1000V / 44.9V = 22.3 \text{ in series}$$

We always round-down when we are figuring out maximum system voltage, so **the answer is B, 22 in series!**

To do this the fast way that we were taught earlier in the book:

$$25C + 32C = 57C$$
$$57C \times 0.0032 = 0.1824$$
$$0.1824 + 1 = 1.1824$$
$$1.1824 \times 38V = 44.9V$$
$$1000V / 44.9V = 22.3$$

22 in series is the answer.

It is good to be able to approach problems differently to completely understand the concepts.

9. There are 2.1 pounds of CO_2 put into the air per kWh of energy produced by coal. You are calculating how much CO_2 you would offset over a 25-year lifespan by installing each solar module. The Imp of the solar module is 8A and the Vmp is 30V. If you install the PV in a location where it will average 1333kWh/kWp/yr during the next 25 years, then how many pounds of CO_2 will you prevent from going into the air for each solar module installed?

    a. 672 pounds
    b. **16,800 pounds**
    c. 10,000 pounds
    d. 33,393 pounds

For this question, we will determine how many kWh each module will make in 25 years and then we will convert the kWh to CO_2 offset.

To determine the power of each module, we can multiply the maximum power voltage by the maximum power current, which is:

$$30V \times 8A = 240W \text{ solar module}$$

We now need to convert the PV module power to kW:

$$240W / 1000W \text{ per kW} = 0.24 \text{ kW per module}$$

Now we can multiply by kWh/kWp/yr to get our annual kWh production for each module:

$$1333kWh/kWp/yr \times 0.24kW = 320kW \text{ per year production}$$

Now we remember to multiply by 25 years of production:

$$320kWh \text{ per year} \times 25 \text{ years} = 8000kWh \text{ per module per 25 years}$$

Now we convert the 8000kWh to CO_2 offset from coal:

$$8000kWh \times 2.1 \text{ pounds of } CO_2 \text{ per kWh} = 16,800 \text{ pounds of } CO_2$$

The answer is B, we offset 16,800 pounds of CO_2 by installing a solar module!

10. A 100MW solar project is expected to make 90% of its year 1 production on year 20. The first year's production came out to be 1492MWh/MW/yr. What is the expected year 20 income if the FIT rate is 10 cents per kWh?

    a. **$13,428,000**
    b. $14,920,000
    c. $149,200,000
    d. $134,280,000

First, we determine 100% of the year 1 energy and then reduce it appropriately:

$$100MW \times 1492MWh/MWp/yr = 149,200MWh \text{ per year}$$

On year 20 we will get 90% of the annual production, so we will multiply our annual year 1 production by 0.9:

$$149,200MWh/yr \times 0.9 = 134,280MWh \text{ per year on year 20}$$

Now we can convert to kWh/yr:

$$134,280MWh/yr \times 1000kWh/MWh = 134,280,000kWh/yr$$

and now convert to dollars:

$$134,280,000 \text{kWh/yr} \times \$0.10/\text{kWh} = \$13,428,000$$

The final answer is A, the system would produce $13,428,000 on year 20.

We could do this on a calculator without paper if we are sharp, rested and fully practiced:

$$100\text{MW} \times 1492\text{MWh/MWp/yr} \times 0.9 \text{ derating} \times$$
$$1000\text{MW/kW} \times \$0.1/\text{kWh} = \$13,428,000$$

I know of a system that is about 100MW in Canada that is making 4.4 times this amount with a FIT that was put in place in 2010 and will last until 2030!

A FIT is a feed-in tariff, which was developed in Germany and is typically a fixed rate for electricity produced by a PV system for 20 years. The rates used to be higher to give people an incentive to go solar when PV systems were much more expensive than they are now.

11. A main service panel is rated for 125A and the main breaker is rated for 100A. If it is a 240V ac service, what is the largest inverter of the following that may be connected with a supply-side connection?

   a. 5kW
   b. 10kW
   c. 15kW
   d. **17kW**

A supply-side connection is dependent upon the size of the service entrance conductors. The service entrance conductors have to have an ampacity that is at least as much as the main breaker.

The sum of all the overcurrent devices of power production sources shall not exceed the rating of the service. This means that we can add more solar than we would ever need to under a net-metering situation.

Since the main breaker is rated for 100A, then let's say the service is at least 100A, so we can calculate the solar breaker to be at least 100A.

A 100A solar breaker would have to be sized to be at least 125% of the current of the inverter when calculating for continuous current:

$$100\text{A} / 1.25 = 80\text{A}$$

or inversely, 100A × 0.8 = 80A (easier to multiply by 0.8 in your head).

An 80A solar breaker at 240V ac when calculating for power is:

$$80A × 240V = 19,200W = 19.2kW$$

Since we can have an inverter that is at least 19.2kW, then the largest of the inverters listed was the largest inverter, which was a **17kW inverter, so D is the correct answer**.

12. A 255W monocrystalline module is used for a 2MW system. What would be the expected performance of the array in 30 years if the degradation rate were 0.5% per year? The system on year 1 made 1452kWh/kWp/yr.

    a. 2490MWh/MWp/yr

    b. 25,0000kWh/yr

    c. **2500MWh/yr**

    d. 7600MWh/yr

First, we will determine how much energy the array made in year 1 by multiplying the array size by the production factor. To make it easier, we will convert kWh/kWp/yr to MWh/MWp/yr, which is the same number, since kWh/kW = MWh/MW.

$$1452kWh/kWp/yr = 1452MWh/MWp/yr$$
$$1452MWh/MWp/yr × 2MW = 2904MWh \text{ per year on year 1}$$

Now, using the compounding equation, we will determine the factor by which to multiply 2904MWh/yr.

If we lose 0.5% per year, then how much do we keep?

$$100\% - 0.5\% = 99.5\% \text{ is how much we keep}$$

Then, we need to turn the percentage into a decimal:

$$99.5\% / 100\% = 0.995$$

Then we need to multiply 0.995 by itself 30 times for 30 years; we can do this by using the X^y button:

$$0.995^{30} = 0.86$$

This means we will lose 14%, or we will keep 86%.

Multiply 0.86 by year 1 production for our answer:

$$0.86 \times 2904\text{MWh/MWp/yr} = 2497\text{MWh per year}$$

The correct answer is C, 2500MWh per year, which is close and rounded off to two significant digits.

13. Which system will perform better, assuming that there is no shading and the orientation of each system is the same?

   a. Flush roof mount
   b. Ballasted roof mount without wind deflectors and extra ballast
   c. Ballasted roof mount with wind deflectors
   d. **Ground mount**

The main difference between these systems is that there are different amounts of airflow around each prospective array, and with more airflow comes more heat dissipation into the air and less heat build-up at the modules.

A flush-mounted roof array or a ballasted array with wind deflectors would have less airflow and would heat up more. The reason we have wind deflectors is to keep the PV from blowing off of the roof without requiring more ballast (weight) to keep the system on the roof. The reason we avoid using too much weight is primarily to keep the roof from collapsing, especially when there is an additional snow load. Without wind deflectors, more ballast blocks are required. It will also increase the number of ballast blocks you would need to buy.

The most airflow would be on a ground mount. Ground mounts are most exposed to the air and will send more heat into the atmosphere than the other types of rooftop systems mentioned. A pole mount will have even more airflow and work slightly better than a ground mount.

It is interesting that many residential rooftop systems are mounted very close to the roof and some with skirts that prevent airflow from coming up under the front of the array. Preventing the airflow with a skirt will prevent flames from coming up under the array if there is a fire and will make the array more fire safe. Many people also think that the skirt makes the array look better. The skirts are common with railless systems, which are popular.

The correct answer is D, ground mount.

14. A PV module measures 1678mm × 998mm and will produce 255W at STC. Assuming full coverage of the roof area, about how much PV could fit on 23m² of rooftop?

a. **3.2kW**
b. 35,000W
c. 28kW
d. 2.8kW

First, we will figure out how many watts per square meter the PV will cover. We will convert mm to m by moving the decimal three places to the left or by dividing by 1000, since there are 1000mm per meter.

$$1678mm / 1m \text{ per } 1000mm = 1.678m$$
$$998mm / 1m \text{ per } 1000mm = 0.998m$$

Then multiply length by width to get area:

$$1.678m × 0.998m = 1.675 \text{ square meters}$$

Now we can determine watts per square meter by dividing the watts of the PV by the square meters of the PV:

$$255W / 1.675 \text{ square meters} = 152.2W \text{ per square meter}$$

Then, if we have 23 square meters to fill up with PV, multiply watts per square meter by square meters:

$$152.2W \text{ per square meter} × 21 \text{ square meters} = 3198W$$

Convert watts to kW:

$$3198W/1000W \text{ per } kW = 3.2kW$$

The correct answer is A, 3.2kW of PV will fit on 21 square meters of roof.

If we wanted to determine the efficiency of the PV, which is important, then we can compare the 152.2W per square meter of production at STC to 1000W per square meter of peak sun testing the modules when the 152.2W per square meter comes out of the PV:

$$152.2W \text{ per square meter} / 1000W \text{ per square meter} = 0.152$$

And then to turn 0.152 into a percentage, we can multiply by 100%:

$$0.152 \times 100\% = 15.2\% \text{ efficient PV}$$

15. Who interprets the National Electric Code?

   a. Contractor
   b. State licensing board
   c. Federal licensing board
   d. **AHJ**

The correct answer is D, the AHJ. The authority having jurisdiction (usually the city or the county building department official) will have the final say in interpreting the Code or even adding requirements that are not covered in the Code.

A contractor will try their best to interpret the Code and have past experiences that will help her or him understand the Code, but **it will be the AHJ who has the final say, so D is the correct answer**.

Usually the state or the federal government will not get involved; however, in some unusual cases, such as on federal or state property, the state or federal government will be the AHJ, which still makes AHJ the best answer.

16. What is the least expensive way to properly connect a 4kW PV array to a house with a 100A panelboard and a 100A main breaker? Assume that there are 100A of loads on the busbar, but there are still extra spaces available.

   a. 4kW inverter load-side connection
   b. **3.8kW inverter load-side connection**
   c. 4kW inverter supply-side connection
   d. 4kW inverter line-side tap

It is very common that there are more load breakers on the busbar than the busbar is rated for. That is why we have a main breaker to protect the busbar, so having 100A of load breakers on a 100A busbar is not a problem. If we had 60A of load breakers on the 100A busbar, we could then add a 40A solar breaker on the load side according to the 2014 NEC, but not according to the 2011 NEC. Check which version of the NEC your state is using.

First, we will calculate the largest inverter we can connect with a load-side connection using the 120% rule.

The 120% rule states that the sum of 125% of the inverter current plus the main breaker cannot exceed 120% of the rating of the busbar. It also states that if we exceed 100% of the busbar rating with the sum of 125% of the inverter current plus the main breaker, then we will have to place the inverter breaker on the opposite side of the busbar from the main supply breaker.

To put it more simply with an equation:

$$\text{(inverter current} \times 1.25) + \text{main breaker} \leq \text{busbar} \times 1.2$$

Now to put in the numbers:

$$\text{(inverter current} \times 1.25) + 100A \leq 100A \times 1.2$$

Now to do the math:

$$\text{(inverter current} \times 1.25) + 100A \leq 120A$$
$$\text{inverter current} \times 1.25 \leq 120A - 100A$$
$$\text{inverter current} \times 1.25 \leq 20A$$
$$\text{inverter current} \leq 20A / 1.25$$
$$\text{inverter current} \leq 16A$$

So, the most the inverter current can be on the load side in this instance is 16A.

Since the voltage for connecting a typical grid-tied inverter to a house in North America is 240V, then we will calculate power:

$$\text{volts} \times \text{amps} = \text{power}$$
$$240V \times 16A = 3840W$$

So the largest load-side inverter that we can connect on the load side is going to be a 3840W inverter, which can also be rounded off to a 3.8kW inverter.

3.8kW inverters are very common for the reason that we did this calculation. They are more common than 4kW inverters. (Be careful of tricky situations where the voltage is 208V. In this case, we would not be able to connect a 3.8kW inverter on this busbar due to the increased current required: 3800W / 208V = 18.3A.)

Any experienced solar salesperson should know that you will often put more PV than inverter capacity on a job. In fact, if we put 4kW of PV on a 3.8kW inverter, the inverter would most likely never in its lifetime power at 3.8kW. We have to remember that the PV was rated at STC and that it would be very unusual for there to be 1000W/square meter of irradiance in the plane of the array and for the PV at the same time to be 25C. Also, besides rarely hitting STC conditions (if ever), the system will also have derating factors, which will also reduce the output from PV to array, such as module degradation, inverter inefficiency, wire losses, soiling, etc. 4kW of PV can by all means be connected to a 3.8kW inverter.

The PV to inverter ratio in this situation would be:

$$4kW \ PV \ / \ 3.8kW \ inverter = 1.05:1$$

In many systems, experts will use a ratio of 1.2:1 to 1.5:1.

With a 1.5:1 ratio, inverter clipping can be expected, but the ac side of the system will be capitalized.

For the answers to this problem: a 4kW inverter will not go on a load-side connection, so answer A is out of the question. A 3.8kW inverter will go on a load-side connection, so B is the best answer for a "least expensive way to properly connect" this system.

Supply-side connections are good, but are usually more expensive, since typically we will have to connect to service entrance conductors by pulling the utility meter. Also, when connecting to service entrance conductors, we are exposed and not protected from high currents coming from the utility. With a load-side connection in this case, the 100A main breaker is easily shut off to make the connection and the 100A main breaker will protect us from dangerous utility currents.

Another thing about the answers to this question is that a line-side tap is a slang term for a supply-side connection. Since we cannot have two correct answers, then neither one can be right.

The correct answer in this case is B, a 3.8kW inverter load-side connection.

17. An industrial building has a utility bill for $100,000 per year for energy, $1,200 for meter charges and $33,000 per year for demand charges. You are planning to install a net-metered PV system without energy storage. If the customer uses 701,000kWh per year, which is the most that you could reduce the customer's bill?

   a. $100,000
   b. $133,000

    c. $134,200

    d. $89,000

In most places, demand charges will not be offset much with a PV system. Demand charges are charges for peak power usage from the utility and it is very difficult for a customer to only use power when the sun is shining and the PV system is producing. In the PV industry it is not ethical to say we are going to reduce demand charges significantly without using an energy storage system.

Metering charges are not typically offset by PV systems, but there was no answer that was for offsetting the energy charges plus the metering charge.

Although we were given the annual kWh usage, it was not used and only a distraction for solving this simple problem.

The correct answer is A, $100,000, since we are only offsetting the energy charges.

It could be argued by someone that their utility will pay them a nominal rate for exporting more energy than they use. This would be very unusual and not feasible in places where people will get a nominal few cents per kWh for exporting.

The best answer of those given is A.

18. There are many PV systems that are mounted close to flat (zero-degree tilt angle). If you were going to mount a system flat, what would be an important distinction to make when purchasing equipment for the flat-mounted system that would give you better performance, that would not give significantly better performance with a tilted array?

    a.   Ballasted system

    b.   **PV laminates**

    c.   Framed modules

    d.   Ungrounded inverter

An ungrounded inverter is typically more efficient than a grounded inverter; however, this is not a benefit exclusive to flat-mounted arrays and will have equal benefits to flat or tilted arrays.

A ballasted system would have the same performance benefits for a flat-mounted array as for a tilted array. The flat-mounted array may require less ballast, since the wind will be less likely to come up from underneath the PV and blow it off the roof.

Framed PV modules would be a draw back and reduce performance for the flat-mounted array. When there are framed modules that are tilted to low or no tilt, there will be soiling build-up. In places where it rains a lot, it would help wash off the soiling.

A PV laminate is a PV module without frames; **B, PV laminates is the correct answer**. With a flat-mounted array, PV laminates would have a performance benefit.

The main drawback for a PV laminate is that there are no sturdy frames with which to mount the PV modules. Frameless modules (laminates) are famous for having a higher breakage rate. The systems with which to mount the laminates typically have special gaskets, so that the PV does not break. PV laminates have been known to break when something hard taps their edge, and when any PV module breaks it is an all-or-none scenario due to the tempered glass. You never just see one crack on PV glass.

19. You are using a subpanel to combine inverter output circuits. The subpanel is rated for 125A. What are the maximum inverter currents that could combine on the subpanel? Also consider that there will be a single 15A load breaker on the subpanel used for a monitoring device.

   a. 40A
   b. **88A**
   c. 25A
   d. 45A

There is a load-side connection rule that requires a label that explains the rule. The required sign says:

> # WARNING:
>
> THIS EQUIPMENT FED BY MULTIPLE SOURCES.
> TOTAL RATING OF ALL OVERCURRENT DEVICES,
> EXCLUDING THE MAIN SUPPLY DEVICE,
> SHALL NOT EXCEED THE AMPACITY OF THE BUSBAR.

This means that if we have a 125A busbar, we could have up to 125A of breakers on the busbar, including solar and load. If there were no load breaker, we could have 125A of solar breakers on the busbar. Since we have a single 15A load breaker for the monitoring device, then we can do the math:

125A busbar – 15A load breaker = 110A available space for solar breakers.
Since the maximum inverter current that a breaker can carry is 80%
of the rating of the breaker, then 80% of 110A is 88A.

This was a new addition to the NEC in 2014. Before the 2014 Code we had to comply with the 120% rule and, given the information in this question, we could not even answer the 120% rule since we were not given the size of the main breaker.

The correct answer is B, 88A of solar breakers.

20. For a stand-alone system that will be used year round, what would be the best tilt angle at 40 degrees latitude, in a location that has no significant shading? You would like the system to have significant production year-round.

    a.   40-degree tilt
    b.   25-degree tilt
    c.   30-degree tilt
    d.   **55-degree tilt**

Since this is a stand-alone system, we would like to get through every season of the year as best we can. In most places, the toughest time of year to get production is the time around the winter solstice. If we can maximize the months surrounding the winter solstice, then we should be fine for the rest of the year.

On the winter solstice, the sun is 23.5 degrees (the tilt of the Earth) below where the array would be facing at a latitude tilt.

There is a general rule of thumb that solar installers have used for decades to optimize production for different seasons:

Annual production	Latitude tilt
Summer production	Latitude –15 degrees tilt
Winter production	Latitude +15 degrees tilt

With the advent of software and advanced production modeling, it has been shown that for annual production often less than latitude tilt is better, since with a less than latitude tilt we can optimize summertime sun better. For a grid-tied net-metering system, a 30-degree tilt is very good for most of the continental USA.

Most off-grid installers would tilt their arrays at latitude +15 degrees to optimize for the wintertime.

Latitude 40 degrees + 15 degrees = 55-degree tilt

The correct answer is D, 55-degree tilt.

21. If you are installing a small PV system on a roof with many different angles, which would be the best solution? The customer also plans on buying an electric car in two years and expanding the PV system.

   a.   Central inverter
   b.   String inverter
   c.   **Power optimizer/inverter combination**
   d.   Battery inverter

The best answer of the above is C, power optimizer/inverter combination. With a power optimizer, there can be multiple orientations and it is not a problem to add PV modules that are dissimilar. A power optimizer will maximum power point track each module individually.

The other inverters cannot have multiple orientations within a single string and it is not as easy to add more modules later for when the electric car is purchased and the system is upgraded.

If a microinverter were a possible answer, it would also be a good answer. It would be hard to differentiate between which would be better. A microinverter is more expensive to buy, but does not need to have a wall-mounted inverter and is more scalable than the inverter with the power optimizers. With power optimizers, you would have to add another wall-mounted inverter at some point, when the ratio of PV to inverter was more than ideal.

22. What is closest to the slope of a 4:12 roof?

   a.   15 degrees
   b.   **18 degrees**
   c.   20 degrees
   d.   22 degrees

To find the slope of a roof, we can either memorize the common roof slope angles or we can do some trigonometry. It is fairly easy to memorize the simple functions to do roof slope angles.

With a roof slope, we often do the ratio of rise to run (rise:run). We also take the run to be 12. We can say 12 inches, but it really does not matter, because it is all just a ratio: 4:12 inches will be the same slope and ratio as 4:12cm.

The rise is the vertical elevation and the run is the horizontal dimension. If we have a triangle and we wanted to find the slope angle of the roof, the rise would be the opposite side of the triangle and the run would be the adjacent side of the triangle.

Recall our trigonometry functions:

tangent = opposite / adjacent

So we can divide 4 by 12 and get 0.333.

If we are to arrive at an angle from the relationship of two sides of a triangle, then we need to use the inverse tangent function (Tan^{-1}).

With 0.333 on our calculator, we press the buttons to get Tan^{-1} and we will end up with the roof slope of 18 degrees.

So **the correct answer is B, 18 degrees**.

With the Casio fx260 calculator, to get the inverse tangent function, press shift and then Tan.

Here are the buttons that we push consecutively:

- 4
- /
- 12
- =
- shift
- tan

Let's do this for all of the roof slopes up to a 12:12 roof:

1 / 12 = shift tan	5-degree slope
2 / 12 = shift tan	9-degree slope
3 / 12 = shift tan	14-degree slope

4 / 12 = shift tan	18-degree slope
5 / 12 = shift tan	23-degree slope
6 / 12 = shift tan	27-degree slope
7 / 12 = shift tan	30-degree slope
8 / 12 = shift tan	34-degree slope
9 / 12 = shift tan	37-degree slope
10 / 12 = shift tan	40-degree slope
11 / 12 = shift tan	43-degree slope
12 / 12 = shift tan	45-degree slope

Practice this on your own in your spare time.

In the case above we rounded off the slope angles to the nearest degree. Being more than 1 degree accurate is not very measurable for a contractor or solar salesperson.

23. According to the 120% rule, what is the largest inverter of the following that can be installed on a 120/240V building with a 225A busbar and a 175A main breaker?

   a. 15kW
   b. 7kW
   c. 8kW
   d. 18kW

The 120% rule states that 125% of the inverter current plus the main breaker cannot exceed 120% of the busbar rating. Also, the 120% rule requires that the solar breaker(s) be placed on the opposite side of the busbar from the main supply breaker.

This time we will solve the problem one step at a time.

120% of the busbar step:

$$1.2 \times 225A = 270A$$

Subtract the main breaker step:

$$270A - 175A = 95A$$

Since 95A is 125% of the inverter current, we can divide 95A to get our inverter current:

$$95A / 1.25 = 76A$$

76A would be the maximum inverter size and we can figure out our maximum inverter size by multiplying by the voltage of 240V:

$$76A \times 240V = 18,240W$$

We can also turn watts into kW by moving the decimal three places:

$$18,240W / 1000 = 18kW$$

Therefore, **the correct answer is D, 18kW.**

24. An investment tax credit is better financially for the system owner if compared to another system composed of the same equipment if. . .?

    a. The system produces 50% more energy every year
    b. **The system costs 50% more**
    c. The system does not break on year 1
    d. The system is installed in a cooler location

An investment tax credit (ITC) is an incentive that is based on the investment made in the PV system. The ITC is a percentage of the price of the system and is not related to system performance. A performance tax credit (PTC) is based on performance and was popular with wind energy.

A performance based incentive gives installers more of an incentive to install a system in a manner that will make as much energy as possible.

Of all of the questions above, the only **answer that was based on the price of the system and not performance was B, the system costs 50% more, which is the correct answer.**

The IRS audited some solar companies that were inflating their prices to get a larger investment tax credit. In these cases the installing solar company still owned the system and sold the power to the customer through a power purchase agreement. They were incentivized to make the price of the system that was reported to the IRS as high as possible.

If I could sell myself a 1W PV system for a million dollars and get a 30% tax credit that I could sell on Wall Street, I could almost retire.

25. What happens with a typical dc-coupled grid-tied battery backup PV system when the grid is working and the batteries are charged?

    a. The charge controller will feed the grid
    b. A backup circuit will switch on and feed the loads with the PV system

    c.   The batteries will feed the grid

    d.   **The diversion load will be the grid**

With a typical dc coupled PV system, under normal circumstances when the batteries are charged and the grid is connected, the inverter will feed the grid as a diversion load. A diversion load is when the batteries are fully charged and the PV system sends the excess power somewhere else.

With a pure stand-alone system, when the batteries are charged and the system is making more power than the loads are using, there can be diversion loads that make use of the excess electricity. Some common diversion loads for a typical stand-alone system would be a heating element in the hot water tank or pumping water.

When a multimodal (bimodal) inverter is used for a dc coupled grid-tied battery backup system, there will be a grid connected output of the inverter for feeding the grid, like a normal utility interactive inverter, and another stand-alone output for the backed-up loads that will work when the grid is down.

The best answer for this problem is D, the diversion load will be the grid, since that is directly the case. It is also the inverter that directly feeds the grid, not the charge controller or the batteries.

26.   You are measuring azimuth with a magnetic compass in California and the magnetic declination is 14 degrees; what would be the true azimuth of the white, magnetic south needle of the compass?

    a.   180 degrees

    b.   0 degrees

    c.   **194 degrees**

    d.   166 degrees

A magnetic compass will point to the magnetic north pole, which moves around over long periods of time and is approximately to the north of the middle of Canada. The true north pole is the axis around which the Earth spins, and is the north that we see on most maps, digital maps and just about everything besides a magnetic compass. We want to align our solar systems with the true north pole, since it is what determines where the sun rises.

There are a few different ways of thinking about and correcting for magnetic declination:

    A. The positive and negative method.

    B. The declination to the east and west method.

    C. The Canada owns the north pole method.

A. Magnetic declination on the West Coast of North America is positive; the East Coast is negative. The line of magnetic declination where there is no needed correction is approximately on the Mississippi River. East of the Mississippi we will subtract the magnetic declination to get the true azimuth and west of the Mississippi we will add the magnetic declination to get true azimuth. To answer our question, since the magnetic declination was a positive number (it will be a negative number on the east coast) we will add to the magnetic azimuth. Since **180 degrees + 14 degrees = 194 degrees, then C is the right answer**.

B. The declination to the east and the west method means that when you are on the West Coast the north needle of your compass will be pointing to the east, so we would call this case magnetic declination to the east by 14 degrees. Since the north pole of the compass is pointing 14 degrees to the east, then the south needle of the compass would be pointing 14 degrees to the west. Since the magnetic south needle of the compass is pointing 14 degrees to the west, we would need to add 14 degrees to account for going west by adding numbers to the azimuth. With this method, keep in mind the rotation of where the compass is going. With magnetic declination to the east, we will be adding numbers.

C. The Canada owns the north pole method is a good way to double-check your work. Since the magnetic north pole is north of the middle of Canada, then we can see in our mind that on the West Coast of the USA the entire compass has shifted 14 degrees clockwise in order to point at Canada instead of the real "true" North Pole. This way we will see that we have to rotate counterclockwise to correct for magnetic declination when we are on the West Coast when switching from magnetic to true azimuth, or we can rotate clockwise if we were going from true azimuth to magnetic azimuth. When the compass shows 180 degrees magnetic azimuth, we can rotate counter-clockwise so that the 180 will be where it truly belongs and what was 194 azimuth on the dial will be shifted 14 degrees, giving us 194 degrees true azimuth.

As you can see, converting between true and magnetic azimuth can make someone dizzy, especially during an exam or on a rooftop in the sunlight. In a real sales situation, you will probably have Google Earth or an equivalent map, which is always going to give you true azimuth.

27. You are looking to install an ungrounded string inverter with two separate MPPs. The inverter will take between 5 and 14 modules in series for the particular temperature conditions and the specified monocrystalline 265W modules. You would like to put two strings of ten modules in series on the roof, but your sales team is worried about a chimney that will shade one PV module for about two hours per day. When that module is totally shaded, about how much total production will be lost?

   a. 5%
   b. 50%
   c. 10%
   d. 25%

Since there are two separate MPPs, the strings will work independently. When one module is shaded, the bypass diodes will activate and bypass the shaded module. With one module bypassed, you will have a string of ten on one MPP and a string of nine on the other MPP. In effect, you will have 19 out of 20 modules working fine. There is a fallacy going around that people think that PV modules are like Christmas lights, and when one module is shaded it will take out the entire string. If we lose 1 out of 20, we can calculate the percentage losses by:

1 module / 20 modules = 0.05 of the modules performance lost
0.05 × 100% = 5% of the power is lost

A is the right answer, since we are only losing 5% of our performance.

As a note, if we lose enough modules to get our voltage below the equivalent of the lowest string size on a hot day, then our performance would suffer more. In this case, we stated that we could have as few as five in series, so nine in series would be no problem.

28. If you were going to check the service voltage on a large industrial building in Nevada, what would you expect the voltage to be?

   a. 120/208V
   b. 120/240V

    c.   **277/480V**

    d.   1000V

The service voltage in most large industrial buildings in the USA is 277/480V three-phase. In the building there are usually transformers that will bring down the voltage to 120/208V so that equipment of lesser voltages can be used.

There are other voltages, and on some very large factories there will be even higher voltages coming from the utility. **The best answer for this question is C, 277/480V.**

If you wanted to check whether numbers matched up with three-phase power, you can multiply the lower voltage by the square root of 3, which is about 1.73, and you will get very close to the higher voltage.

$$120V \times 1.73 = 208V$$
$$277V \times 1.73 = 479.21V, \text{ which is close enough to } 480V$$

In Canada, instead of 277/480V they usually have 347/600V coming into their large buildings:

$$347V \times 1.73 = 600V$$

If you are going to check the voltages on a building, be careful. Be sure to inspect and check your meter first. Also, rather than checking the voltage, often you can see the voltage written somewhere, such as on the meter.

29. What size overcurrent protection device should you have at the output of a 4kW inverter connected to the supply side of the main breaker on a house?

    a.   20A

    b.   **25A**

    c.   30A

    d.   40A

The inverter breaker is sized based on 125% of the inverter current; you then round-up to the next common overcurrent protection device size.

Since we are not given the current of the inverter, like we would usually be on the inverter, we are going to calculate the current of the inverter using the voltage and power of the inverter.

$$\text{voltage} \times \text{current} = \text{power}$$
$$\text{power} / \text{voltage} = \text{current}$$

Since residential voltage in North America is 120/240V (even in Canada), and 99% of the inverters that are installed on houses are 240V, then we will use 240V for the voltage and 4000W for the 4kW inverter.

$$4000W / 240V = 16.7A$$

Then we will multiply by 1.25 to get 125% of the inverter current (if you forget this step, you will size the overcurrent protection wrong):

$$16.7A \times 1.25 = 20.9A$$

Then we round-up to the next common overcurrent protection device size.

For breakers between 10A and 50A, common overcurrent protection device sizes are in 5A increments, so we can round-up to a 25A overcurrent protection device. **The correct answer is B, 25A overcurrent protection device**.

25A breakers are uncommon and many solar installers will get away with using a 30A breaker. 25A is still the right answer.

30. If a 5kW PV system were to last for 50 years and produce an average of 1400kWh/kWp/yr, how much money would the PV system save if the average adjusted for inflation price of a kWh over 50 years is 30 cents per kWh?

    a. $50,000
    b. $350,000
    c. $77,000
    d. **$105,000**

It is reasonable to say that many well-constructed PV systems may last 50 years; however, their inverters will probably need to be changed a few times. After all, most PV is made out of glass, metal and crystal. Many windows last 100 years if there is no vandalism or natural disasters.

To calculate for this problem, first we will determine how much energy the system will make in a single year:

$$5kW \times 1400kWh/kWp/yr = 7000kWh \text{ per year}$$

Then we will multiply by 50 years:

7000kWh/year × 50 years = 350,000kWh per 50 years

Then we will multiply by the price of a kWh adjusted for inflation:

350,000kWh × $0.30 = $105,000

The correct answer is D, $105,000.

Strategy

The NABCEP PV Technical Sales Exam is more technical than sales. It would be most difficult for the exam writers to write sales questions, because much of sales is subjective and the non-subjective sales questions would differ based on states or programs that will change. State questions cannot be in an exam that is written for Americans and Canadians. Program-based questions will change rapidly with the times. Solar finance changes by the year and by the state.

It is recommended to focus on the technical for this exam. I believe that NABCEP would like to see solar salespeople know how to sell solar ethically in order for the solar industry to have a sterling reputation.

Think of the technical dilemma that solar salespeople will find themselves in if they do not properly understand the technical aspects of installing PV, calculating energy production and where the system will be connected.

In the 1970s there were great solar rebates for solar hot water systems, and the bad quality of many of the installations was legendary – so legendary that the solar industry's reputation took decades to recover. It is most unfortunate when there are less than ethical solar installers selling systems that they do not understand and making promises that they cannot keep.

We have recovered as an industry and are the fastest job creator in the new economy. Study hard, pass the exam and do something good for the world.

Here are some tips for a good score:

1. Read this book carefully, especially all of the practice questions.
2. Be familiar with all of the NABCEP Entry Level Material.
3. Study regularly in advance, preferably at the time of day of the exam.
4. Be on a sleep cycle that gets you up two hours before the exam time for a week.

5. Eat a good breakfast.
6. Practice with a Casio fx260 calculator – it's solar powered!
7. Do not consume too much caffeine; nobody falls asleep during an exam.
8. Arrive early and make sure you know where you are going.
9. Do not leave the exam early. Catch your mistakes.
10. Expect to be NABCEP Certified!

RECOMMENDED RESOURCES:

- *Solar Photovoltaic Basics*, by Sean White
- *Solar PV Engineering and Installation*, by Sean White
- National Electric Code
- NABCEP website: www.nabcep.org
- DSIRE website: www.dsireusa.org

If you are limited on study time, studying this book and *Solar Photovoltaic Basics* should be enough to get you through the exam. Make sure to know the information thoroughly. Too much information can be overwhelming, so take in the knowledge one step at a time.

Remember, the most valuable person in a solar organization is a good salesperson. Be good and sell solar. The company cannot survive without you.

Do yourself and the world a favor. Sell a lot of solar!

Best of luck!

Index

A reference in *italics* indicates a figure and references for tables are in **bold**.

AHJ (authority having jurisdiction): defined 7; and interpretation of the National Electrical Code 168; and structure safety decisions 139; *see also* building department
airflow 166
AM (air mass) 4, 39, 40
amp hours and battery capacity 31–2
amps (amperes) 29, 33
array alignment 122–3; *see also* orientation
array fencing 138
automatic transfer switches 11
azimuth: calculation 109–10; defined 9; magnetic declination adjustments and 178–80

ballast blocks 166, 171–2
batteries: battery capacity 31–2; BIPV lithium battery backup systems 63; de-coupled grid-tied battery backup PV systems 177–8; diversion loads 178; energy calculation 32; interactive inverters and 62–3; old-style off-grid system 62, 63; parallel connections 32, *33*; series connections 32, *32*; and utility policies 63

breakers: breaker sizing 104–5; common breaker and inverter combinations **68**; inverter breaker sizing 11; solar breakers 164–5; *see also* main breakers
building department: defined 7; for seismic reinforcement approval 98
busbars: and the 120% rule 66–8; rating and inverter sizing 110–13, 136, 153–4
bypass diodes: in 60 or 70-cell modules 4; defined 4; in PV systems 42--4; and shading *43*, 96, 157, 180

Canada 181
carbon dioxide: offset calculations 107–8, 133, 162–3; offsets and ecologically conscious consumers 133–4
CFL lighting 128
charge controllers 42
cold temperature string sizing 44–9, 106–7, 141, 161–2
combiner boxes: dc disconnects 59; defined 5; fuses 59; in grid-tied PV systems 60; and inverters 59; parallel connections 54
commercial buildings: building department approval and 98; typical voltage of 105–6; *see also* industrial buildings

Printed in the United States
by Baker & Taylor Publisher Services

Printed in the United States
by Baker & Taylor Publisher Services